Lecture Notes in Biomathematics

Managing Editor: S. Levin

51

Oscillations
in Mathematical Biology

Proceedings of a conference
held at Adelphi University, April 19, 1982

Edited by J. P. E. Hodgson

Springer-Verlag
Berlin Heidelberg New York Tokyo 1983

Editor

J. P. E. Hodgson
Department of Mathematics and Computer Science
Adelphi University
Garden City, New York 11530, USA

AMS Subject Classifications (1980): 92-06, 92 A 09, 92 A 40, 58 F 14, 65 L 07

ISBN-13: 978-3-540-12670-6   e-ISBN-13: 978-3-642-46480-5
DOI: 10.1007/978-3-642-46480-5

Introduction

The papers in this volume are based on talks given at a one day conference held on the campus of Adelphi University in April 1982.

The conference was organized with the title "Oscillations in Mathematical Biology;" however the speakers were allowed considerable latitutde in their choice of topics. In the event, the talks all concerned the dynamics of non-linear systems arising in biology so that the conference achieved a good measure of cohesion.

Some of the speakers chose not to submit a manuscript for these proceedings, feeling that their material was too conjectural to be committed to print. Also the paper of Rinzel and Troy is a distillation of the two separate talks that the authors gave. Otherwise the material reproduces the conference proceedings.

The conference was made possible by the generous support of the Office of the Dean of the College of Arts and Sciences at Adelphi. The bulk of the organization of the conference was carried out by Dr. Ronald Grisell whose energy was in large measure responsible for the success of the conference. I should also like to thank the participants for their time and the contributors for all their efforts in preparing their manuscripts.

TABLE OF CONTENTS

*Indicates name of person who gave the talk

Address of Contributors

Professor Gail A. Carpenter
Northeastern University
Mathematics Department
Boston, Massachusetts 02115

Professor G. Bard Ermentrout
Mathematical Research Branch
N.I.H.
Bethesda, Maryland 20205

Professor John Feroe
Vassar College
Department of Mathematics
Poughkeepsie, New York 12601

Professor Stephen Grossberg
Boston University
Boston, Massachusetts 02215

Professor Brian Hassard
State University of New York at Buffalo
Department of Mathematics
106 Diefendorf Hall
Buffalo, New York 14214

Professor Nancy Kopell
Mathematics Department
Northeastern University
Boston, Massachusetts 02115

Professor John Rinzel
Mathematical Research Branch
N.I.H.
Bethesda, Maryland 20205

Professor William Troy
University of Pittsburgh
Department of Mathematics and Statistics
Pittsburgh, Pennsylvania 15261

A ONE-VARIABLE MAP ANALYSIS OF BURSTING

IN THE BELOUSOV-ZHABOTINSKII REACTION *

By

John Rinzel   and   William C. Troy [†]

Mathematical Research Branch
National Institute of Arthritis, Diabetes
and Digestive and Kidney Diseases
National Institutes of Health
Bethesda, Maryland 20205

Department of Mathematics
University of Pittsburgh
Pittsburgh, Pennsylvania 15260

ABSTRACT

We consider a three-variable model for the Belousov-Zhabotinskii reaction run
in a continuous flow stirred tank reactor (CSTR). We focus on the complex oscillation (burst) patterns for relatively low flow rates. From the chemical model we
derive a one-variable discontinuous mapping for the concentration of a critical
species from one spike to the next. We consider a reasonable piecewise linear
approximation to the map whose solution behavior is described analytically. Various
features (e.g. bistability) of the map solutions predict burst patterns for the
continuous model.

* Also to appear in: Proc. 1982 AMS Summer Res. Conf.: Nonlinear Partial
                      Differential Equations (J.A. Smoller, ed.), Durham, N.H.

† Research supported in part by: NIHRCDA K04 NS00306-05 and
                                  NSF Res. Grant No. NCS8002948

1. INTRODUCTION

The Belousov-Zhabotinskii [1, 17, 18] (BZ) reaction is one of the best characterized chemical systems which exhibits both self-sustained oscillations and chemical
excitation as well. The reaction consists of the metal ion oxidation by bromate
ion of easily brominated organic materials. In a batch reactor, when the reagent
is in the oscillatory state, the oscillations are most apparent in the ratio of the
oxidized and reduced forms of the metal ion catalyst. For other recipes, in which
the oscillations are suppressed, the system may exhibit excitability. Recent experiments reveal a wide range of possible behavior when the BZ reaction is run in a CSTR
(continuously stirred tank reactor). In addition to regular periodic oscillations at
some flow rates [2,4,7,10,11,16], the observed phenomena include: bursts of oscillations in which one or more nearly identical pulses are separated by more or less
regular intervals of quiescence [2, 7, 13], multiple steady states [2, 7], and irregular patterns such as quasi-periodic and chaotic oscillations [5, 10, 11, 16]. Different approaches have been followed to model the CSTR system [6, 8, 11, 12, 14, 15].
Here, we focus our attention on the five variable model of Janz, Vanecek and Field [6].
Their model is based on an irreversible, batch reactor Oregonator [3] with flow terms,
and with instantaneous dependence of the stoichiometric parameter f on the brominated
organic substance. Recently, Rinzel and Troy [9] have simplified the five variable
model into a system of three equations, namely

$$(1.1) \qquad \dot{y} = \frac{1}{s} (-y - g(y,r)y + f(p)z) + \frac{r}{s} (y^0-y)$$

$$\dot{z} = w(g(y,r) - z) + \frac{r}{s} (z^0-z)$$

$$\dot{p} = \frac{1}{s} (y + 2g(y,r)y + \frac{q}{2} [g(y,r)]^2 - f(p)z) + \frac{r}{s} (p^0-p)$$

where

$$(1.2) \qquad g(y,r) = (1-y-r/s^2 + [(1-y-r/s^2)^2 + 4q(y + rx^0/s^2)]^{1/2})/2q ,$$

$$f(p) \equiv Fp^2/(K\bar{p}^2 + p^2) .$$

Here $"\cdot" = \frac{d}{dt}$, $y \propto Y = [Br^-]$, $z \propto Z = [M^{+(n+1)}]$ (the higher oxidation state of the metal ion catalyst), $p \propto P =$ the concentration of the brominated derivative of the organic substrate. Also, y^0, z^0 and p^0 are proportional to the concentrations of Y, Z and P in the feed stream. Other parameter values are s = 77.27, w = .161, q = 8.375 x 10^{-6}, F = 4.0, K = .0005 and $\bar{p}$ = (1/3) x 10^7. The parameter r is proportional to the flow rate and it is the easiest physical parameter to tune in the CSTR system.

Our computations show that Eqns. (1.1)-(1.2) exhibit qualitatively different oscil-latory responses for high, low or intermediate flow rates. Relatively small values of r lead to bursts of several oxidation pulses separated by an interval of quiescence (IQ) during which the system is near a steady state of reduction. Such a pattern (see Fig. 1) corresponds to bursts computed in [6], also [15, (Fig. 5)], and observed experimentally in [2, 7, 13]. Bursting at relatively larger r (and K), is charac-terized by single reduction pulses separated by quiet periods of high oxidation. These solutions share qualitative features with calculated ones of Showalter et al. [12], and experimental ones of Schmitz et al. [11], except that the IQ's in these cases exhibit noticeable small oscillations. For intermediate r (and K) we find regularly repetitive pulses of large amplitude with no identifiable IQ between pulses. The system tends to a steady state for extremely high or low r. Experimental responses exhibit a similar qualitative dependence on flow rate [4, 10]: regular oscillations for intermediate r with complex oscillation patterns between this range and very high or very low flows where there is steady state behavior; in some cases, small amplitude oscillations are found between the steady state and complex oscillation regimes.

Our goal here is to understand low flow rate bursting patterns of the model (1.1)-(1.2) similar to that shown in Figure 1. For this it is convenient to view the CSTR as a "batch reactor" subsystem (i.e. the y and z equations) coupled with

the dynamics for the stoichiometric parameter f. As we see in Fig. 1B there is
a small net change in f from one spike to the next. Thus we are led naturally
to a one dimensional mapping equation for f which describes the net change in f
per spike. We compute the map numerically and derive a piecewise linear approximation. In our analysis of the piecewise linear approximation we examine the
behavior of solutions as a function of the flow rate. In agreement with solutions
to (1.1)-(1.2) we find that the number of pulses per burst increases with r. Moreover, we find that there are r-intervals in which the map exhibits two different
stable bursting patterns. This predicts, and we have verified, the simultaneous
existence of two stable bursting solutions in the continuous system (see Fig. 6).
For some parameter ranges, the map exhibits aperiodic behavior as the flow rate
increases. Further studies are now underway to determine the behavior of the continuous system at these higher values.

In Section 2 we analyze the batch reactor subsystem and develop the one-dimensional (1-D) mapping equation. Following that, in Section 3, we describe and
analyze the bursting patterns predicted from the 1-D map. Section 4 is discussion.

2.   THE BATCH REACTOR AND A ONE DIMENSIONAL MAPPING EQUATION

As was noted in the Introduction, we find it convenient to view the CSTR as a
batch reactor subsystem coupled with the dynamics for the stoichiometric parameter
f . The batch reactor subsystem consists of the equations

$$(2.1) \qquad \dot{y} = \frac{1}{s} (-y - g(y,0)y + fz)$$

$$\dot{z} = w(g(y,0) - z)$$

where, from (1.2),

$$g(y,0) = (1-y + [(1-y)^2 + 4qy]^{1/2})/2q .$$

Here f is held constant and the flow terms are ignored (r = 0). Rinzel and Troy
[9] have numerically shown (Fig. 2) that there is a Hopf bifurcation of unstable

periodic orbits in Eqn. (2.1) which occurs as follows: there are values $f_H \approx .5$ and $f^H \approx 1.52$ such that if $f\varepsilon(f_H, f^H)$ then Eqn. (2.1) has an unstable steady state solution (denoted by $(y_0, z_0) = (y_0(f), z_0(f))$) surrounded by a large amplitude stable periodic orbit (of relaxation type). As f passes through f_H (from above) or f^H (from below) there occurs a Hopf bifurcation of small amplitude unstable periodic solutions from the steady state. On intervals of the form $(f_H - \mu, f_H)$ and $(f^H, f^H + \eta)$ there coexist a large amplitude stable periodic solution, a smaller unstable periodic orbit and a locally stable steady state. For $f > f^H$ the two families of periodic orbits coalesce at a value $f_\nu \approx 1.58$ called the "knee" of the bifurcation diagram. For $f > f_\nu$ the steady state is globally stable and no periodic solutions exist. The bifurcation diagram has the form shown in Fig. 2.

To obtain a one dimensional simplification of the CSTR equations (1.1)-(1.2) we find it useful to interpret the bursting solution (Fig. 1) in terms of the batch reactor bifurcation diagram (Fig. 2). Initially for Fig. 1, $f(0) = 1.2$. Thus, $f_H < f(0) < f^H$ and the (y,z) subsystem immediately enters the spiking mode. During each large cycle in (y,z) space f undergoes a corresponding small increase Δf. We let f* denote the critical value of f such that if $f(0) = f*$ at the start of a spike then $\Delta f = f_\nu - f*$. After a sufficient number of spikes, f enters the interval $(f*, f_\nu)$. Subsequently, one more large oscillation in (y,z) space causes f to exceed f_ν. At this point (y,z) quickly approaches the stable steady state of the batch system and the large oscillations are extinguished. The p dynamics now cause f to decrease while (y,z) remains close to the steady state. This period is referred to as the interval of quiescence (IQ) between bursts. It persists until f dips below f^H and the steady state of the batch system becomes unstable. Rinzel and Troy [9, pp. 1782-1784] show that in fact f must decrease sufficiently below f*, to a value approximately equal to 1.43, before the large oscillations in (y,z) space are rekindled (Fig. 1B). Again, each of these spikes causes a net increase in f and the entire loop described above is repeated (see schematic representation of this loop in Fig. 2).

We now derive the one dimensional mapping. We first note that during the spiking mode the trajectories traced out in (y,z) space are nearly identical [9], independent of f. Thus, whenever y is at its peak value during a spike we compute the corresponding value of f. This leads to a discrete sequence $\{f_n\}_{n \in Z}$ of f values. We seek to determine the functional relationship $f_{n+1} = \phi(f_n)$. The periodic bursting pattern of Fig. 1 only gives four points. We determine $\phi(f)$ for other values of f as follows. First, choose f and compute the large amplitude periodic solution of the batch system. Let $\bar{y}$ denote the maximum value of y, and $\bar{z}$ the corresponding value of z. Next, determine from Eqn. (1.2) the value of $\bar{p}$ which corresponds to $\bar{f}$, i.e. such that $f(\bar{p}) = \bar{f}$. Finally, let $(y(0),z(0),p(0)) = (\bar{y},\bar{z},\bar{p})$ in the full system, Eqns. (1.1)-(1.2). At the next spike in y the corresponding value of f is $\phi(f)$. Our numerics show that to a good approximation $\phi(f)$ is linearly increasing for f < f*, and $\phi(f)$ approaches f_ν as f tends to f* from below (see cresses of Fig. 3). If f* < f < f_ν then recall (Fig. 1B) that the next value of f where a spike occurs is f ≈ 1.43 < f*. Our numerics show that $\phi(f)$ is discontinuous at f = f*, and monotonically decreasing for f* < f < f_ν (again, see the crosses). As a simplification we approximate $\phi(f)$ by the discontinuous piecewise linear function in Fig. 3. This map also has a periodic solution; each of the four points in this discrete periodic solution corresponds to a pulse.

3.   BURST PATTERNS FOR THE 1-D MAP

In the preceding section we developed a discrete mapping $f_{n+1} = \phi(f_n)$. Here we restrict attention to the piecewise linear approximation

$$(3.1) \qquad \phi(f) = f_\nu + m(f-f^*) , \quad f < f^*$$

$$= f_a + (\frac{f_b-f_a}{f^*-f_\nu})(f-f_\nu) , \quad f^* \le f \le f_\nu .$$

Figure 4 empirically justifies using the discrete map to describe the bursting behavior of the CSTR model, (1.1)-(1.2), for r = 0.00414. To apply this description for other flow rates requires that we make explicit the assumed dependence of ϕ upon r. Certain features may be deduced from the parametric dependence of bursts upon r as reported in Section 3.A2 of [9]. There we found that during the spiking phase the net increment of f per spike decreases for larger flow rate. From this it follows that, as r increases, the left segment of ϕ(i.e. for f < f*) moves closer to the 1:1 line and f* is nearer to f_ν. We make the simplest assumption by supposing that m is independent of r and f* increases with r. Since we have not yet computed the exact dependence of f* on r we shall here study the map behavior as a function of f* to qualitatively mimic dependence of the CSTR responses upon r.

Next we consider the right segment of ϕ(i.e., for f > f*). From our analysis of the IQ [9], this segment must have negative slope for each r. That is, the further f is above f^H at the start of the IQ, the further f falls below f^H before the next spike. Our numerical experiments to date indicate a small relative variation of f_a and f_b with respect to r (e.g., $d(\ln f_a)/dr$). Initially, we suppose f_a and f_b are independent of r but later we shall consider weakening this restriction. We summarize the above assumptions on (3.1) as follows:

  i)   f* increases with r,

  ii) m is independent of f*, m < 1,

  iii) f_a is independent of f*,

  iv) f_b is independent of f*.

Figure 4 illustrates periodic solutions of the map (3.1) for several values of f*. An N-point discrete solution corresponds to an N-pulse burst for the continuous model, (1.1)-(1.2). Consistent with our findings [9] for (1.1)-(1.2), the number of pulses per burst increases with flow rate, i.e. with f*.

Since the approximate map is piecewise linear we may calculate explicitly the discrete periodic orbits shown in Fig. 4. Let a periodic solution of N points be represented by $\{f_1, \ldots, f_N\}$ where $f^* \leq f_N \leq f_\nu$. Then, from (3.1), we have

(3.2a) $f_1 = f_a + (f_N - f_\nu)(\frac{f_b - f_a}{f^* - f_\nu})$

and

(3.2b) $f_n = f_\nu + m(f_{n+1} - f^*)$

$$= \sum_{j=0}^{n-2}(f_\nu - mf^*) + m^{n-1}f_1 \quad, \quad 2 \le n \le N \ .$$

Now evaluating (3.2b) for n = N and using (3.2a) we obtain a linear equation for f_N in terms of N and f*:

(3.3) $f_N = f_a + m^{N-1}(\frac{f_b - f_a}{f^* - f_\nu})(f_N - f_\nu)$

$$+ (\frac{1 - m^{N-1}}{1-m})\ [f_\nu - f_a - m(f^* - f_a)] \ .$$

This equation has a unique solution f_N, however only for appropriate ranges of N and f* is the solution f_N in the interval $[f^*, f_\nu]$ so that it corresponds to a periodic burst pattern. It is not difficult to show that in this range f_N is a decreasing function of f*. Thus, for a given N, the N-point periodic solution first appears as f* increases through $f^*_{N,a}$ at which value f_N enters the interval $[f^*, f_\nu]$ from above. Thus from (3.3) $f^*_{N,a}$ is defined by the condition $f_N = f_\nu$:

(3.4) $f^*_{N,a} = \dfrac{(m^{N-1} - m^{N-2})f_a + (m^{N-2} - 1)f_\nu}{m^{N-1} - 1}$.

The N-point periodic solution then disappears for f* = $f^*_{N,d}$ ($f^*_{N,d} > f^*_{N,a}$) when f_N disappears into the map's discontinuity at f*. Thus we have

(3.5) $f^*_{N,d} = \dfrac{(m^N - m^{N-1})f_b + (m^{N-1} - 1)f_\nu}{m^N - 1}$.

For this piecewise linear map we can also determine the stability of the N-point periodic solution. Namely, the solution is stable if the multiplier σ_N, given by

(3.6) $\sigma_N = m^{N-1}(f_b - f_a)/(f^* - f_\nu)$,

satisfies $|\sigma_N|<1$; the solution is unstable if $|\sigma_N|>1$. Since $\sigma_N < 0$ and σ_N decreases with f*, we find instability for $f* > f^*_{N,\nu}$ where $\sigma_N = -1$ for $f* = f^*_{N,\nu}$, i.e. from (3.6)

$$f^*_{N,\nu} = f_\nu - m^{N-1}(f_b-f_a) .$$

In Fig. 5 we indicate (for m, f_a, f_b as in Fig. 3) the intervals of f* over which an N-point periodic solution exists. The vertical bar represents the value of $f^*_{N,\nu}$ if it falls in the interval $(f^*_{N,a}, f^*_{N,d})$; to the right of the bar the N-point solution is unstable. For the assumptions and parameter values of this figure, all N-point periodic solutions for $2 \leq N \leq 5$ are stable while all those for $N \geq 9$ are unstable.

An immediately noticeable feature of Fig. 5 is that more than one periodic solution may exist for certain intervals of f*. One may see this analytically by comparing (3.4) and (3.5) to find $f^*_{N,a} < f^*_{N-1,d}$. We emphasize that such coexistence is a qualitative consequence of the right branch of ϕ having negative slope. Because periodic solutions for low values of N are stable, these overlapping intervals of existence imply bistability (of periodic solutions) for certain ranges of f*. For example, for f* = 1.5413 (the vertical arrow in Fig. 5), both a stable 3-point and stable 4-point periodic solution coexist. These two solutions are shown in the upper two panels of Fig. 6. The value of f_N for the N = 3 orbit lies to the left of f_N for the N = 4 orbit. With respect to initial values in the interval $(f*,f_\nu)$, each orbit's domain of attraction is an interval about its respective f_N value. These two domains of attraction are separated by the intersection of the dotted path (traced backwards from the point (f*,f*)) with the right branch of ϕ. For initial values less than f*, the domains of attraction are alternating bands.

The 1-D map thus makes a definite prediction for coexistence and hysteresis of N-pulse periodic solutions for the continuous model. The lower panels of Fig. 6 verify this prediction. They show that for flow rate, r = 0.003750, the Eqns. (1.1)-(1.2) have both a stable 3-pulse pattern and a stable 4-pulse pattern. The initial

conditions were chosen consistent with the domains of attraction for the 1-D map
solutions. We have also seen coexistence of solutions to the continuous model
for some other values of r (e.g., near the r-value at which an N-pulse solution
disappears).

Returning to Fig. 5, one sees for increasing f* how solutions progress from
an N-point to an (N+1)-point periodic pattern. One should ask however what the
observed discrete pattern is when f* exceeds the value $f^*_{8,\nu}$, 1.5634..., beyond
which no stable N-point periodic solutions exist. In this situation only unstable
periodic orbits exist and moreover there are at least two of them for each such f*.
The map response is then aperiodic mixing among patterns which have different numbers
of points on the left branch corresponding to the different coexistent unstable,
N-point, periodic solutions. For example, from Fig. 5, if f* satisfies $f^*_{8,d}$ < f* <
$f^*_{11,a}$ ($<f^*_{10,d}$) then the response would be aperiodic mixing between 9- and 10-point
patterns. To see this, we follow a suggestion by J. Keener (private communication)
and consider the return map for ϕ on the interval (f*,f_ν). This return map is also
a discontinuous, piecewise linear function. Each segment has negative slope, intersects the 1:1 line, and corresponds to a pattern with a distinct number of points
on the left branch of ϕ. The aperiodic mixing behavior for f* large enough corresponds to iterating between the various different segments of this return map.

One wonders if such aperiodic behavior, e.g. mixing between N- and (N+1)-pulse
bursts, can be found for the continuous CSTR model. So far we have not observed
such aperiodic solutions to Eqns. (1.1)-(1.2). We have computed stable periodic
burst patterns with as many as 28 pulses per burst.

On the other hand, it is not requisite that the 1-D map exhibits aperiodic
behavior for large N; whether or not it does, depends on the parameters. In particular, if the lower branch is not too steep, i.e. if

$$(3.7) \qquad \frac{f_b - f_a}{f_\nu - f_a} < \frac{1-m}{2-m} \quad ,$$

then each N-point periodic solution is stable and the map response is always periodic. For a slowly increasing flow rate one would observe an orderly insertion of an additional spike (because $f^*_{N+1,d} > f^*_{N,d}$) into the periodic burst pattern. Note, the inequality (3.7) follows from (3.5) and (3.6) by requiring for all N that $\sigma_N > -1$ for $f^* \leq f^*_{N,d}$. The case of Fig. 4 however falls into the parameter regime

$$(3.8) \qquad \frac{f_b - f_a}{f_\nu - f_a} > \frac{1-m}{m}$$

for which all N-point periodic solutions are unstable for large enough N. In this case, the inequality (3.8) follows from (3.4) and (3.6) and the demand that $\sigma_N < -1$ for $f^* = f^*_{N,a}$ when N is large enough.

To correlate further the 1-D map predictions with solutions to (1.1)-(1.2) we should consider more carefully the dependence of the map parameters on the flow rate r (or f^*). For example, we might relax assumptions iii) and iv) and allow f_a and f_b to depend on f^*. If f_b were decreased with f^* appropriately the map would exhibit only stable periodic solutions.

4. DISCUSSION

We have considered a simplified three variable model, Eqns. (1.1)-(1.2), for the BZ chemical system in a CSTR. We focus on the complex oscillation (burst) patterns for relatively low flow rate (Fig. 1). For these bursting responses there is a small positive increment in the species P (and therefore f) during each spike of the burst and then a slow substantial decrease between bursts. By interpreting such patterns and the dynamics of f in terms of the batch reactor subsystem and its bifurcation diagram (Fig. 2) we derive a one variable discontinuous mapping for the value of f from one spike to the next (Fig. 3). We consider a reasonable piecewise linear approximation to the map whose solution behavior is described analytically.

Various features of the map solutions mimic, and moreover predict, burst patterns for the continuous model. For example, the number of spikes per burst increases by one as the flow rate r increases through critical values (Figs. 4 and 5). For r above the low flow rate regime, the discrete and continuous models both exhibit regular (non-bursting) oscillations; for the map this occurs as f* increases above f_ν and a stable fixed point appears. In addition, the 1-D map predicts hysteresis between the N-pulse and (N+1)-pulse patterns as r is tuned; i.e. for a certain interval of r values these periodic solutions coexist. This prediction is a qualitative one (depending primarily on the negative slope of the map's right branch) and has been verified (in several cases) for the continuous model (Fig. 6). Depending on the map parameters and their assumed dependence on r, the map response may become aperiodic for large enough r. In this case the map predicts aperiodic mixing between bursts with different numbers of pulses. Such behavior has not yet been observed for the continuous model. On the other hand, for different map parameter assumptions all the N-point periodic solutions are stable and this predicts a regular progression from the N-pulse to the (N+1)-pulse periodic burst as flow rate increases. We expect to examine other parameter ranges and assumptions for both models to check for further, and perhaps improved, comparability.

We also seek comparison with experimental data. For example, the models treated here predict hysteresis and bistability which might be investigated experimentally. Experimental work is very active but there is disparity between the different chemical recipes and flow rate regimes which the different groups consider [10]. Our bursts resemble those in [2, 7, 13]. Extensive low flow rate experiments have been done by the Texas group [16] (also, some in Bordeaux: see [10]). Their observed complex oscillation patterns are somewhat different from that in Fig. 1. Their patterns are similar to high flow rate patterns in experiments of [5,11] but the qualitative dependence on r is reversed. In our terminology, their burst has a single spike and the IQ exhibits small oscillations. As r increases the number of small cycles between spikes decreases until a periodic response of only large spikes is observed (this is the intermediate flow rate regime). The progression from one periodic pattern to the next, e.g. one with N+1 small cycles to one with N small cycles, is not sharp (as in Fig. 5). Rather the experimental data indicate transition regions where aperiodic mixing between the two patterns occurs. There is some evidence that period-doubling cascades lead to these chaotic regimes. Many of these data have been interpreted in terms of 1-D maps obtained from experimental data [10, 14, 16] or differential equation models without chemical interpretation [8].

A discontinuous map of the form we study (but not necessarily piecewise linear) may also exhibit period-doubling behavior and chaos. With curvature in the map one expects a cascade of period-doubling bifurcations subsequent to the destabilization of an N-point periodic solution. However, we view this as fine-structured chaos characterized by irregular timing of pulses in N-pulse bursts. We distinguish this from the chaos of aperiodic mixing between N and (N+1)-point patterns which we have described above. This latter behavior occurs when the return map of the interval $[f^*, f_\nu]$ has multiple branches and more than one branch is expanding, i.e. has slope less than minus one. In this case, the unstable fixed point of each such branch acts as a repeller. A typical trajectory spirals outward on one branch, jumps to a different branch where it spirals out and then is rein-jected back to the first branch.

FIGURE LEGENDS

FIGURE 1.   Bursts of oxidation pulses (from [9]). Solution, y vs. t and f vs. t,
of simplified CSTR model, Eqns. (1.1)-(1.2) for r = 0.00414. After initial transient
stage the response tends to a periodic pattern of four-pulse bursts separated by
regular intervals of quiescence (IQ). Initial conditions are y(0) = 0.5, z(0) = 11.0,
f(p(0)) = 1.2. The dashed horizontal line in B is for $f = f^H \approx 1.52$ (see Section 2).

FIGURE 2.   The bifurcation of small amplitude unstable periodic orbits of the batch
reactor subsystem, Eqns. (2.1), occurs at $f_H \approx 0.5$ and $f^H \approx 1.52$. The knee of the
curve occurs at $f_\nu \approx 1.579$. The path with arrows represents the hysteresis loop
followed by the bursting solution of Fig. 1 for r = 0.00414. While following the
upper branch of the curve the solution is in the spiking mode. After falling off
the knee the solution follows the f-axis while (y,z) is near the batch steady state
during the IQ. The value f*, 1.544, refers to the discrete map of Fig. 3.

FIGURE 3.   The piecewise linear approximation (solid) for the discontinuous map
$f_{n+1} = \phi(f_n)$ of f from one spike to the next. The data points (6 above and 5 below
the 1:1 line), shown as crosses, were determined by solving the CSTR model, Eqns.
(1.1)-(1.2), for different initial conditions (see text). Vertical arrows indicate
the periodic, four pulse, burst pattern of Fig. 1. The piecewise linear map also
has a four-point periodic solution for these map parameters: f* = 1.544, f_ν = 1.579,
f_a = 1.436, f_b = 1.46 and slope .94 for 1.42 ≤ f < f* .

FIGURE 4.   Discrete stable periodic solutions to map Eqn. (3.1) for different values
of f*, a parameter corresponding to flow rate of the CSTR. Parameter values f_ν, f_a,
f_b, and m are as in Fig. 3. An N-point discrete solution should correspond to an
N-pulse periodic solution to the continuous model, Eqns. (1.1)-(1.2). Here observe
that N increases with f*.

FIGURE 5.   Intervals of existence, $f^*_{N,a} \leq f^* \leq f^*_{N,d}$, for N-point periodic solution to map Eqn. (3.1). Parameter values f_ν, f_a, f_b, and m are as in Fig. 3. Periodic solutions with N ≤ 5 are stable; those with N ≥ 9 are unstable. Solutions with 6 ≤ N ≤ 8 are stable for f^* to left of vertical bar (at $f^* = f^*_{N,\nu}$) and unstable to right. For f^* = 1.5413 (vertical arrow) both three-point and four-point stable orbits exist.

FIGURE 6.   For f^* = 1.5413 the map Eqn. (3.1) has both a three and four-point stable periodic solution as shown in the upper two panels. The dotted path leading from (f^*,f^*) divides the interval (f^*,f_ν) into two subintervals which define the domains of attraction for the two periodic orbits. For r = 0.00375 Eqns. (1.1)-(1.2) exhibit the corresponding coexistence of stable three and four-pulse periodic bursting solutions. The lower four panels show the y and f components of these solutions.

REFERENCES

1. Belousov, B. P., A periodic reaction and its mechanism, Ref. Radiats. Med., 1958 (1959) 145- .

2. De Kepper, P., A. Rossi, and A. Pacault, Etude experimentale d'une reaction chimique periodique. Diagramme d'etat de la reaction de Belousov-Zhabotinskii, C. R. Acad. Sci. Ser. C 283 (1976) 371-375.

3. Field, R. J. and R. M. Noyes, Oscillations in chemical systems. IV. Limit cycle behavior in a model of a real chemical reaction, J. Chem. Phys., 60 (1974) 1877-1884.

4. Graziani, K. R., J. L. Hudson, and R. A. Schmitz, The Belousov-Zhabotinskii reaction in a continuous flow reactor, Chem. Eng. J., 12 (1976) 9-21.

5. Hudson, J. L., M. Hart, and D. Marinko, An experimental study of multiple peak periodic and nonperiodic oscillations in the Belousov-Zhabotinskii reaction, J. Chem. Phys. 71 (1979) 1601-1606.

6. Janz, R. D., D. J. Vanecek, and R. J. Field, Composite double oscillation in a modified version of the Oregonator model of the Belousov-Zhabotinskii reaction, J. Chem. Phys., 73 (1980) 3132-3138.

7. Marek, M. and E. Svoboda, Nonlinear phenomena in oscillatory systems of homogeneous reactions - experimental observations, Biophys. Chem., 3 (1975) 263-273.

8. Pikovsky, A. S., A dynamical model for periodic and chaotic oscillations in the Belousov-Zhabotinskii reaction, Phys. Lett., 85A (1981) 13-16.

9. Rinzel, J. and W. C. Troy, Bursting phenomena in a simplified Oregonator flow system model, J. Chem. Phys., 76 (1982) 1775-1789.

10. Roux, J.-C., Experimental studies of bifurcations leading to chaos in the Belousov-Zhabotinskii reaction, Proc. of Los Alamos Conf., Order in Chaos, (1982) to appear in Physica D.

11. Schmitz, R. A., K. R. Graziani, and J. L. Hudson, Experimental evidence of chaotic states in the Belousov-Zhabotinskii reaction, J. Chem. Phys., 67 (1977) 3040-3044.

12. Showalter, K., R. Noyes, and K. Bar-Eli, A modified Oregonator model exhibiting complicated limit cycle behavior in a flow system, J. Chem. Phys., 69 (1978) 2514-2524.

13. Sorensen, P. G., in general discussion, Proc. Faraday Soc. Symp., 9 (1974) 88-89.

14. Tomita, K. and I. Tsuda, Towards the interpretation of Hudson's experiment on the Belousov-Zhabotinskii reaction, Prog. Theoret. Phys., 64 (1980) 1138-1160.

15. Turner, J. S., Periodic and nonperiodic oscillations in the Belousov-Zhabotinskii reaction, Preprint from Discussion Meeting, Kinetics of Physiochemical Oscillations, Aachen, Sept., 1979.

16. Turner, J. S., J.-C. Roux, W. D. McCormick, and H. J. Swinney, Alternating periodic and chaotic regimes in a chemical reaction-experiment and theory, Phys. Lett., 85A (1981) 9-12.

17. Zhabotinskii, A. M., Dokl. Acad. Sci. Nauk SSSR, 157 (1964) 392- .

18. Zhabotinskii, A. M., Periodic course of oxidation of malonic acid in solution (investigation of the kinetics of the reaction of Belousov), Biophysics, 9 (1964) 329-335.

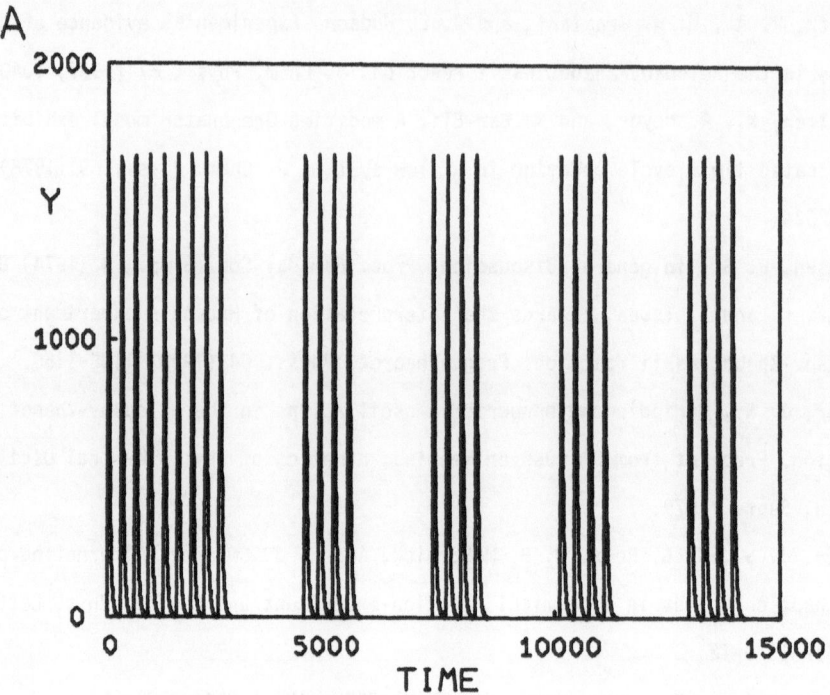

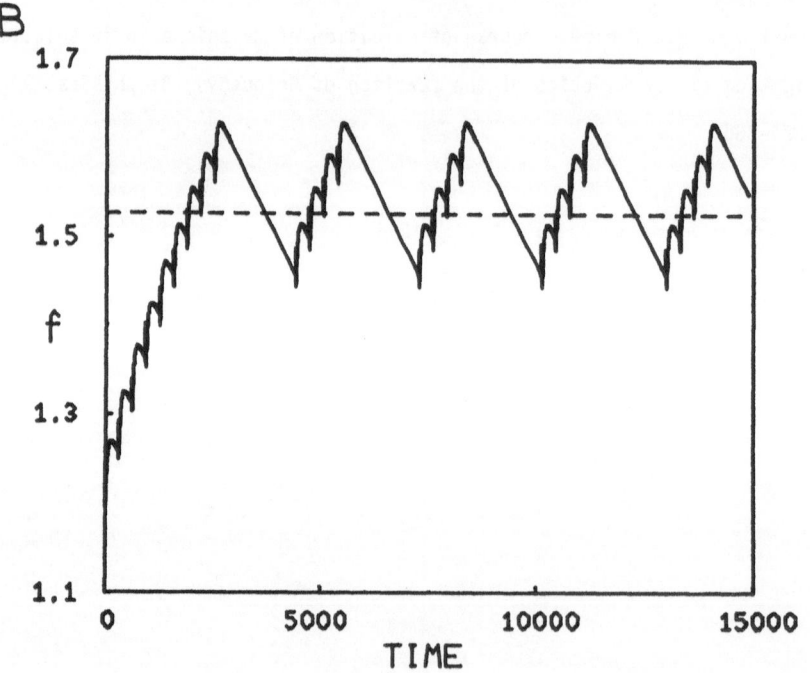

FIGURE 1

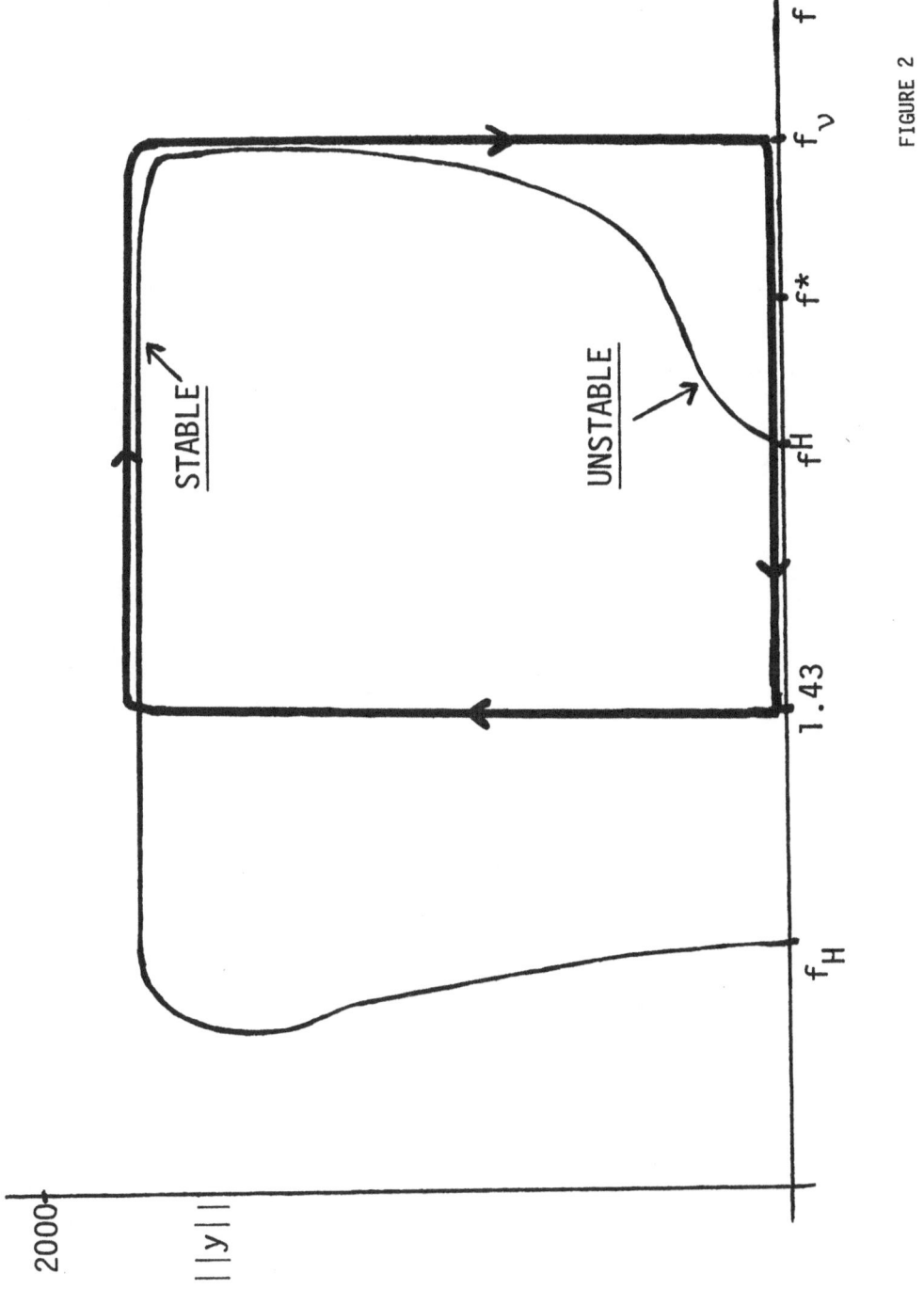

FIGURE 2

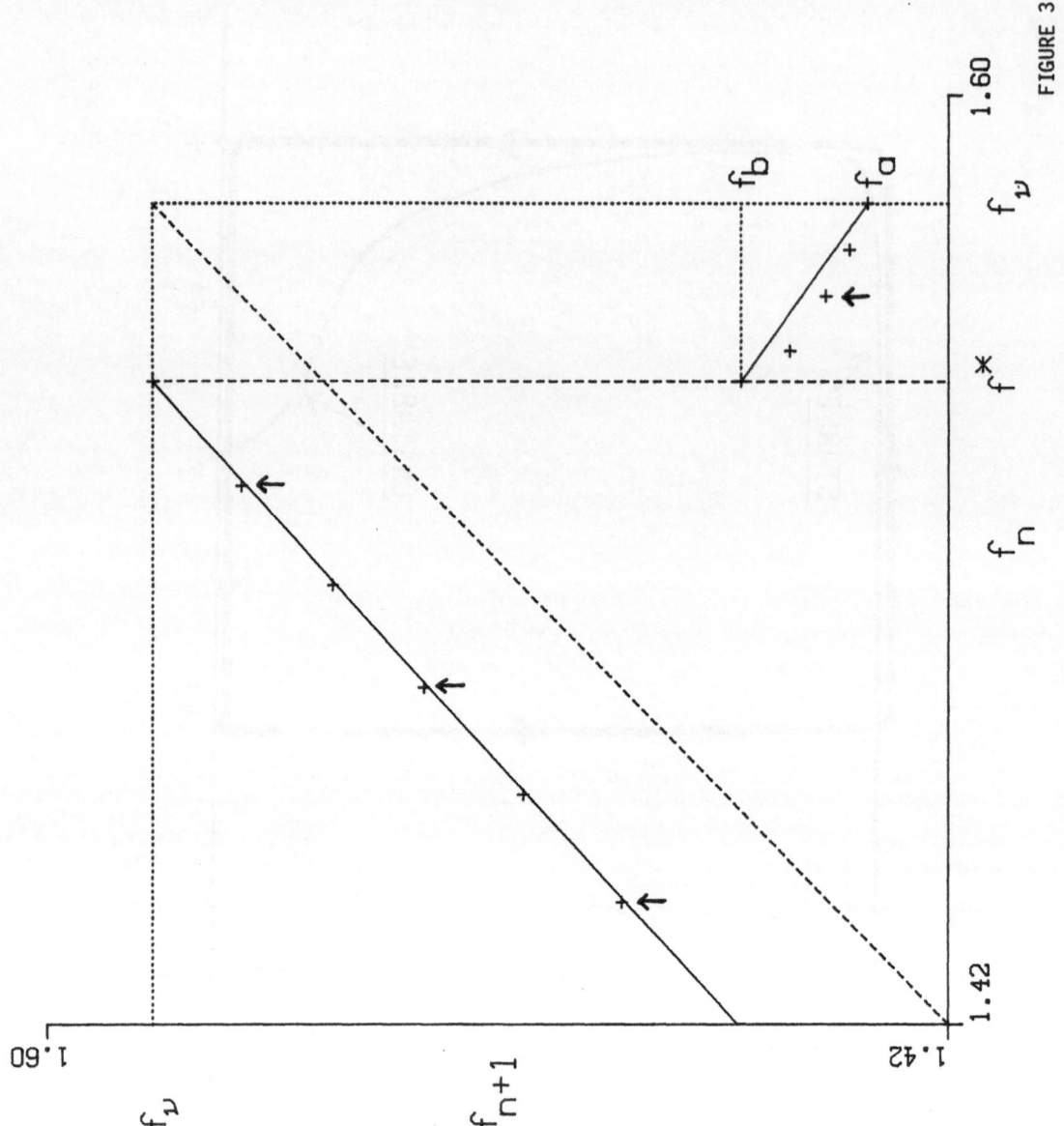

FIGURE 3

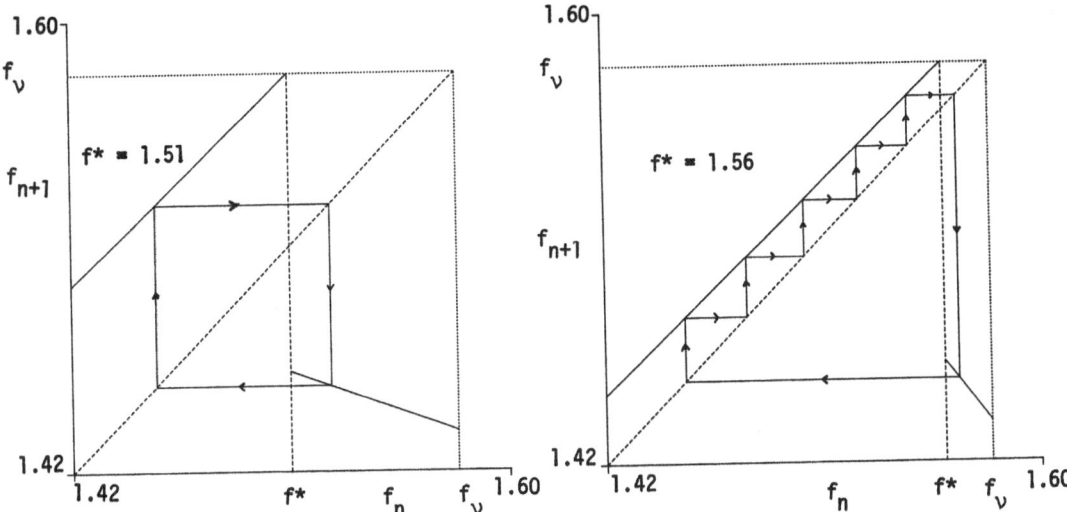

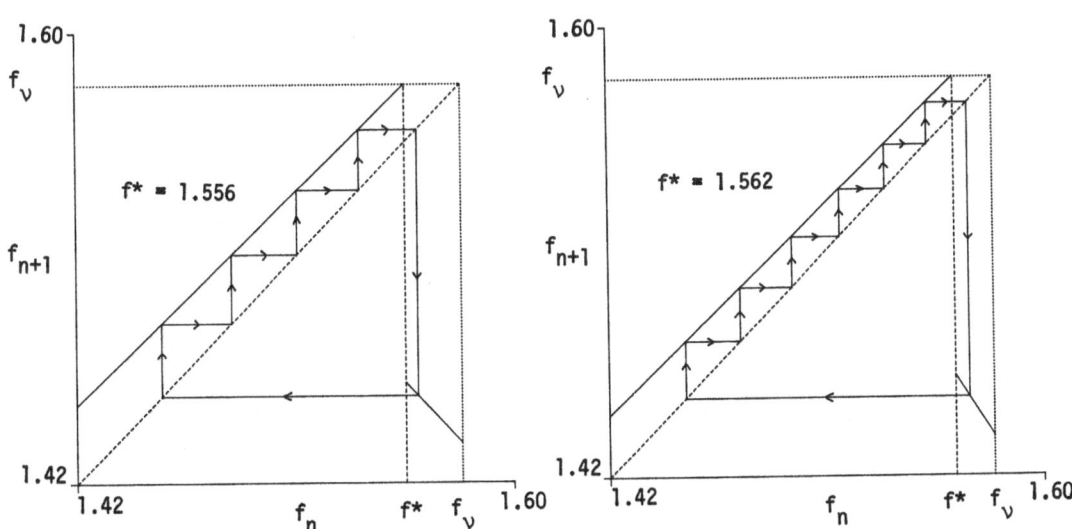

FIGURE 4

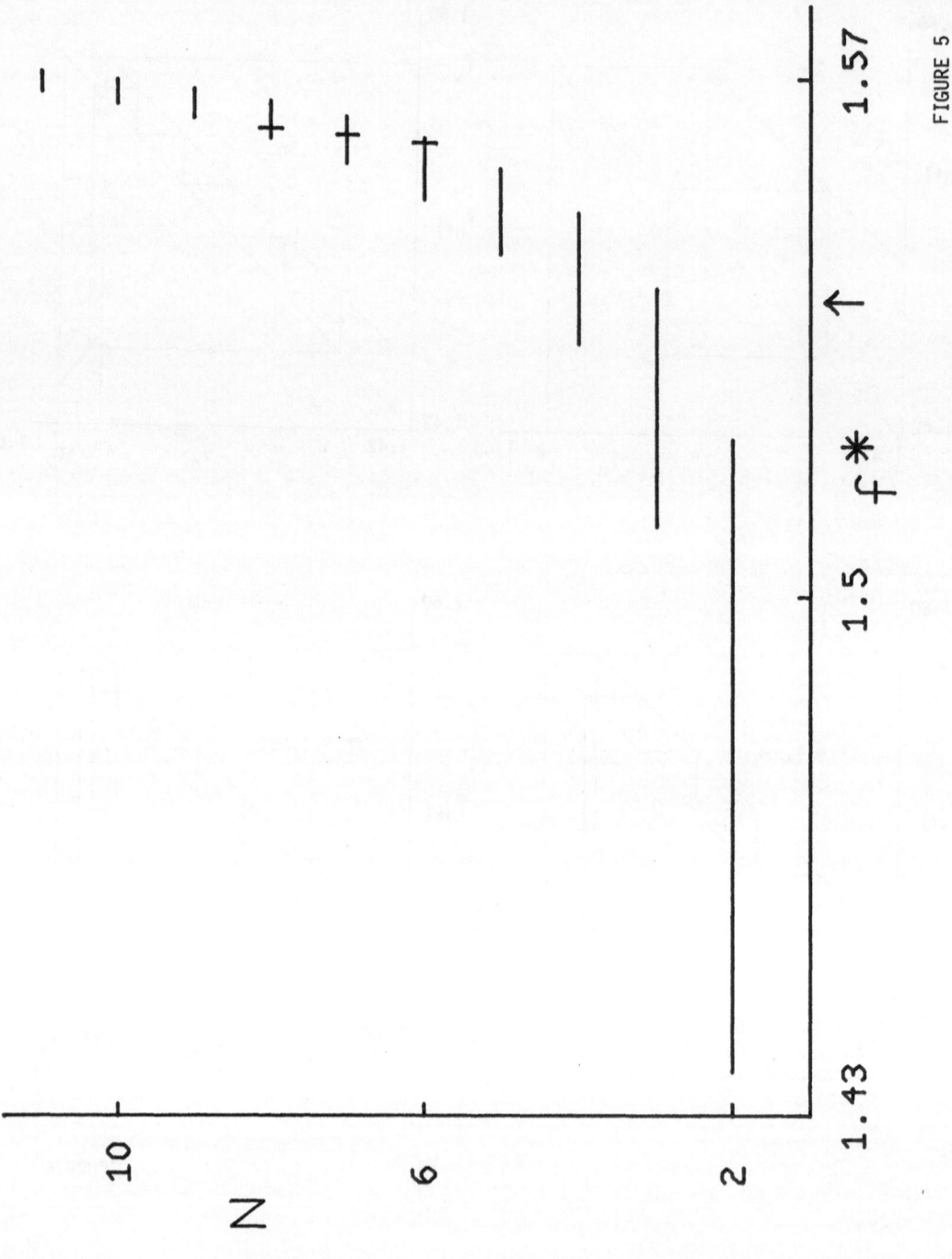

FIGURE 5

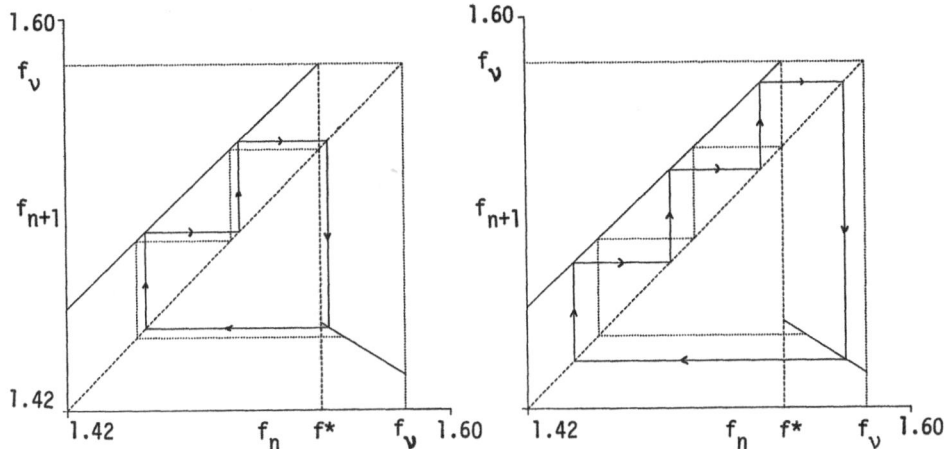

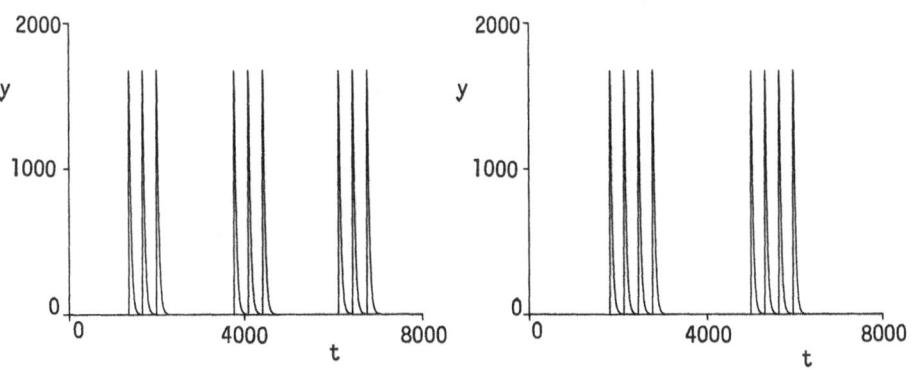

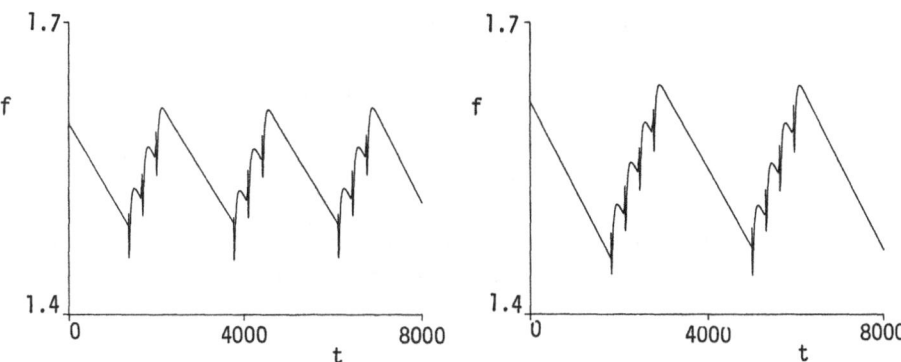

FIGURE 6

Coupled Oscillators and

Mammalian Small Intestines

N. Kopell
Mathematics Department
Northeastern University
Boston, MA 02115

and

G.B. Ermentrout
N.I.H., N.I.A.D.D.K
Mathematical Research Branch
Bethesda, MD 20205

The mathematics to be discussed below was motivated by certain phenomena observed in mammalian small intestines, which consists of layers of smooth muscle fiber. It is known that the muscle fibers support travelling waves of electrical activity which run from the oral to the aboral end [1-4]. These, in turn, trigger waves of muscular contractions [1,2] via high frequency electrical spikes. The spikes, which have much higher frequency, are considered to be consequences of the slow waves, so we shall be concerned only with these slow electrical waves.

If a section of the intestine is sliced into pieces of length 1-3 cms., each piece is capable of supporting spontaneous oscillations at a constant frequency, with a wave form that is close to sinusoidal [4]. (The origin of these oscillations is controversial [2].) Furthermore, over a substantial section of the intestine there is a linear gradient in the frequency of these oscillations, higher in the oral end than in the aboral. In vivo, the measured electrical activity along the (intact) intestine displays frequency plateaus. That is, the slow wave electrical activity has constant frequency over portions of the intestine, with jumps in frequency at places that do not appear to have physiological significance. (Indeed, the jumps are not necessarily at the same place in successive experiments [4].) The frequency on any given plateau is at least as high as the highest of the natural frequencies along that segment.

The slow-wave electrical activity has been modelled by several investigators [5-9] as a chain of loosely coupled oscillators, mostly Van-der-Pol equations with almost-sinusoidal limit cycles. The exact form of the oscillators, the gradient in frequencies, the form and strength of the coupling and the amount of inhomogeneity and/or anisotropy in the coupling vary among those authors. For a variety of related equations, they produced simulations (digital or electronic) which yield frequency plateaus. Ultimately, the intestinal phenomena should probably be understood in terms of a continuum model; we set as our first task to analytically understand the dynamical behavior of a discretized system (as in the above simulations) in a context as free as possible of the (unknown) details of the oscillators and the coupling. Even here, the mathematical questions turn out to be very rich, and this paper provides only a beginning contribution.

One question we asked was: to what extent could the plateau formation be understood in terms of a phase model, i.e., one in which each oscillation is represented by only one dimension, essentially its phase angle. Such an approach to the intestinal problem was suggested by J. Keener [10]. We believe that the phase model does indeed capture the essential behavior, at least for the discretized system, and provided that the coupling is sufficiently weak.

Deriving a phase model

To get such a phase model, one starts with a very general model of n+1 coupled oscillators, in which the coupling is linear but not necessarily isotropic, i.e., the forward and backward "pulls" are not necessarily the same:

$$(1) \qquad X'_k = F_k(X_k) + \epsilon D(X_{k+1} - (1+\alpha)X_k + \alpha X_{k-1})$$

with $X_0 \equiv 0 \equiv X_{n+2}$. Here $\epsilon \ll 1$ and $\alpha-1$ is a measure of the anisotropy. The reaction term of (1),

$$(2) \qquad X'_k = F_k(X_k) \equiv F(X_k) + \epsilon R_k(X_k, \epsilon)$$

is a perturbation, uniform in k, of

$$(3) \qquad X' = F(X) , \quad X \in R^m , \quad F:R^m \to R^m$$

where (3) is assumed to have a stable limit cycle with frequency ω_0. Thus, for ϵ sufficiently small, (2) must also have a limit cycle, with frequency ω_k , where $\omega_k = \omega_0 + O(\epsilon)$. D is an m×m matrix.

One now chooses new coordinates in R^m adapted to certain m-1 dimensional submanifolds of R^m which are known in the context of oscillations as "isochrons" [11]. Using the isochrons, one can find a change of coordinates $X = B(\theta, Y)$ for (3), with $\theta \in S^1$ and $Y \in R^{m-1}$, such that (3) becomes

$$\theta' = \omega_0$$

$$Y' = L(\theta)Y + O(|Y|^2)$$

where $L(\theta)$ is an $(m-1) \times (m-1)$ dimensional matrix and $Y=0$ is the limit cycle [11]. Using the same change of coordinates for (2), we get

$$\theta_k' = \omega_k + \epsilon\bar{R}_k(\theta_k, Y_k, \epsilon)$$
$$Y_k' = L(\theta_k)Y + O(|Y_k|^2, \epsilon)$$

where

$$\int_0^{2\pi} \bar{R}_k(\theta_k, 0, \epsilon)d\theta_k = O(\epsilon).$$

It can then be shown [12] that the $m(n+1)$ dimensional phase-space of (1) has an invariant $(n+1)$ dimensional torus T^{n+1}, and on this manifold the equations take a special form. The existence of the invariant torus is immediate just from the hypothesis that each oscillator (3) has a stable limit cycle, and that the coupling is weak [13,14]. To get the form of the equations, we take $\phi_k \equiv \theta_{k+1} - \theta_k$, $k=1,\ldots,n$ and $S_k \equiv Y_k/\epsilon$. Then, using averaging techniques [15], it can be shown [12] that (1) may be written as

(4a) $$\theta_1' = \omega_1 + \epsilon H(\phi_1) + O(\epsilon^2)$$

(4b) $$\phi_k' = \omega_{k+1} - \omega_k + \epsilon[H(\phi_{k+1}) + \alpha H(-\phi_k) - H(\phi_k) - \alpha H(-\phi_{k-1})] + O(\epsilon^2)$$

$$H(\phi_0) \equiv 0 \equiv H(\phi_{n+1})$$

(4c) $$S_k' = O(1)$$

where the $O(\epsilon^2)$ terms in (4a,4b) depend only on θ_1 and the $\{\phi_k\}$. $H(\phi)$ is a 2π-periodic function. Rescaling time using $\tau = \epsilon t$, up to $O(\epsilon)$ Equation (4b) becomes the phase model equation

(5) $$\dot{\phi}_k = \Delta_k + H(\phi_{k+1}) + \alpha H(-\phi_k) - H(\phi_k) - \alpha H(-\phi_{k-1})$$

with $\dot{\phi}_k \equiv d\phi_k/d\tau$ and $\epsilon\Delta_k \equiv [\omega_{k+1} - \omega_k]$. Note that, to lowest order, the equations for the $\{\phi_k\}$ are independent of θ_1 and form an n-dimensional system; the full equations, of course, need not have an n-dimensional invariant subtorus of T^{n+1}.

It is not apparent from (5), but can be seen from the explicit computations for H, that H depends not only on the behavior of (2) on each cycle, but also on its behavior in a neighborhood of that limit cycle. For example, if (2) is the two-dimensional system

(6)
$$\begin{bmatrix} c_1' \\ c_2' \end{bmatrix} = \begin{bmatrix} \lambda & -\omega \\ \omega & \lambda \end{bmatrix} \begin{bmatrix} c_1 \\ c_2 \end{bmatrix}$$

with $\lambda = 1 - (c_1^2 + c_2^2)$, $\omega = \omega_k + \omega'[1 - (c_1^2 + c_2^2)]$ and $D = \begin{bmatrix} d_1 & d_2 \\ d_3 & d_4 \end{bmatrix}$ then

(7)
$$H(\phi) = A\sin\phi + B[\cos\phi - 1]$$

with

$$A = \left(\frac{d_1 + d_4}{2}\right) + (d_2 - d_3)\frac{\omega'}{4}$$

$$B = (d_1 + d_4)\frac{\omega'}{4} + \left(\frac{d_3 - d_2}{2}\right)$$

Note that the coefficients A and B are influenced by the average of the diagonal entries of D, the lack of symmetry of D, and the number ω' which measures the frequency dependence of (6) on the amplitude $c_1^2 + c_2^2$. The function H combines these influences: to lowest order, the behavior of the system (1) depends only on α and the function H. Similarly, suppose we have a collection of Van-der-Pol oscillators in the nearly sinusoidal regime. The coupled Equations (1) (for $\alpha = 1$) are

(8)
$$\ddot{X}_k + \delta(X_k^2 - 1)\dot{X}_k + (1 + \delta\omega_k)X_k =$$

$$\varepsilon[b(\dot{X}_{k+1} - 2\dot{X}_k + \dot{X}_{k-1}) + c(X_{k+1} - 2X_k + X_{k-1})$$

$$+ d(\ddot{X}_{k+1} - 2\ddot{X}_k - \ddot{X}_{k-1})]$$

where $\delta \ll 1$, $\omega_{k+1} - \omega_k = O(\varepsilon)$, and the terms involving b,c and d represent, respectively, resistive, inductive and capacitive coupling. Then

$$H(\phi) = b\sin\phi + (c - d)[\cos\phi - 1].$$

In this example, there is no frequency dependence on amplitude (to lowest order), and so all the terms of H come from the coupling.

Analysis of the phase model

The n-dimensional phase system (5) is much simpler than (1). Nevertheless, it is still quite complicated. For example, even for H of the form (7), $\alpha=1$, it is quite hard to find the critical points of (5). For much of the rest of the paper, we shall restrict ourselves to the simpler case $H = \sin\phi$, for which the formation of the frequency plateaus can be more easily understood. We will also show why this example is in some sense pathological, and how the qualitative picture changes when H is more complicated. Most of the details can be found in [12].

For H an odd function (for example $H=\sin\phi$) Equation (5) simplifies. Motivated by the intestinal example, for which the natural frequencies form a decreasing linear gradient, we take $[\omega_{k+1} - \omega_k] = -\varepsilon\beta < 0$, (independent of k). Then (5) becomes

$$(9) \qquad \dot{\phi}_k = -\beta + \sin\phi_{k+1} - 2\sin\phi_k + \sin\phi_{k-1}$$

with $\phi_0 \equiv 0 \equiv \phi_{n+1}$. We shall examine how the dynamics of (9) changes as β increases, i.e., as the frequency gradient from the top to the bottom of the chain increases. For convenience, we take n odd, i.e., $n=2j-1$ for some j. (n even leads to a more degenerate situation.)

For β sufficiently small, (9) has a unique stable critical point. Since $\phi_k \equiv \theta_{k+1} - \theta_k$ is the phase difference between successive oscillators, a critical point for (9) corresponds to a phase-locked state, i.e., one for which the phase differences have (to lowest order in ε) a value independent of time, though dependent on k). The critical points of (9) are easily calculated using a vector form of (9):

$$(10) \qquad \dot{\underline{\phi}} = -\beta\underline{\Delta} + K\underline{\sin\phi}$$

where

$$\underline{\phi} = \begin{pmatrix} \phi_1 \\ \vdots \\ \phi_n \end{pmatrix} , \quad \underline{\Delta} = \begin{pmatrix} 1 \\ \vdots \\ 1 \end{pmatrix} , \quad \underline{\sin\phi} = \begin{pmatrix} \sin\phi_1 \\ \vdots \\ \sin\phi_n \end{pmatrix}$$

and K is a tri-diagonal matrix with -2 along diagonal and $+1$ adjacent on either side. It can be seen that for β sufficiently small,

$$-\beta\underline{\Delta} + K\underline{\sin\phi} = 0$$

has 2^n critical points, each having as k^{th} component one of the 2 solutions $\phi_k^{\pm}$ to

$$(11) \qquad \sin\phi = K^{-1}(\beta\underline{\Delta})_k ,$$

where the R.H.S. of (11) denotes the kth entry of the vector $K^{-1}(\beta\underline{\Delta})$, and ϕ_k^- denotes the solution with smaller absolute value. (See Fig. 1.)

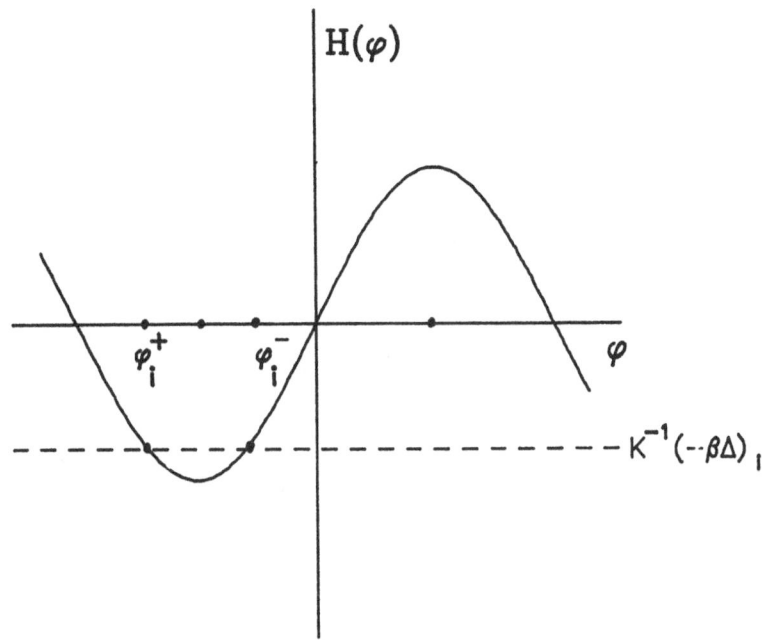

Figure 1
The two possible choices $\phi_i^{\pm}$ for the ith component of a critical point of (9).

It can be shown that exactly one of these critical points is stable. (It is the one for which $\phi_k = \phi_k^-$ ∀k.)

As β increases, the critical points at first simply move closer to one another. The stable critical point $\underline{\xi}_0 = \underline{\xi}_0(\beta)$ moves further from $\underline{\xi}_0(0) = (0,0,0)^t$, with the jth component ξ_{0j} having the largest absolute value. (See Fig. 2). (Recall that $j=(n+1)/2$.) However, at some criti-cal value β_0 of β, all the critical points coalesce in pairs and dis-appear for $\beta > \beta_0$. The sink coalesces with a saddle ($\underline{\xi}_j = \underline{\xi}_j(\beta)$) having all the same component as the sink except for k=j; at β_0, the jth com-ponents of ξ_{0j} and ξ_{jj} also become equal, namely to $-\pi/2$.

For $\beta < \beta_0$, the saddle $\underline{\xi}_j$ has a one-dimensional unstable manifold. For $\beta_0-\beta$ small, $\underline{\xi}_0$ and $\underline{\xi}_j$ are close, and it follows easily from [16] that one branch of this unstable manifold must go to the sink $\underline{\xi}_0$. (See Fig. 3.) What is much harder to show is that the other branch also tends to $\underline{\xi}_0$; furthermore, the two branches, plus the critical points, form a smooth, invariant, attracting circle which is homotopic to the circle

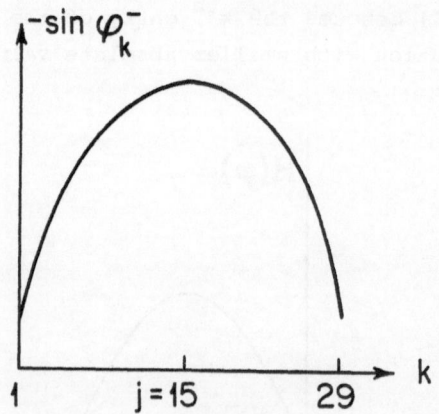

Figure 2
The graph of $-\sin\phi_k^-$ vs. k for the case of isotropic coupling.
$n = 29$, $j \equiv (n+1)/2 = 15$

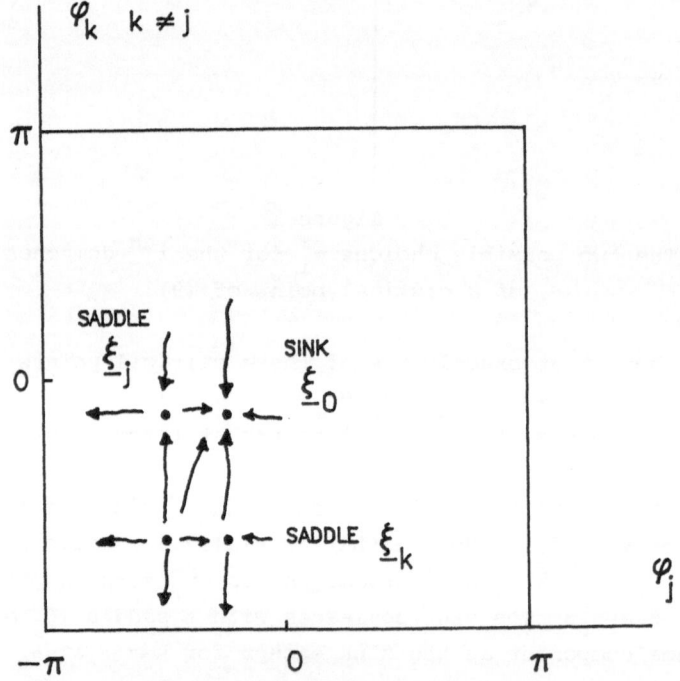

Figure 3
Schematic representation of the dynamics of (9), with a unique sink $\underline{\xi}_0$ and a saddle $\underline{\xi}_j$ which coalesces with $\underline{\xi}_0$ as $\beta \to \beta_0$. The two (one-dimensional) branches of the unstable manifold of $\underline{\xi}_j$ form a smooth invariant circle.

$\phi_k = 0$ $k \neq j$, and $0 \leq \phi_j \leq 2\pi$ (on the n-dimensional torus spanned by $\phi_1, \ldots, \phi_n$ with ϕ_k measured mod 2π). It follows from invariant manifold theory [13,14] that for $\beta > \beta_0$, $\beta-\beta_0$ not too big, there is still an attracting invariant manifold for (5). Note that for $\beta > \beta_0$ this circle contains no critical point, so it must be an attracting limit cycle for (5). The fact that the circle is homotopic to $\phi_k = 0$ $\forall k \neq j$ means that, for $k \neq j$, ϕ_k undergoes "small" oscillations during the limit cycle; only ϕ_j breaks away and traverses its entire range.

The limit cycle corresponds to the existence of a pair of frequency plateaus with a β-dependent jump in frequency between them. This can be seen as follows: The "frequency" of an oscillator coupled to others re-quires a definition, and we suggest using

$$(12) \qquad \lim_{T \to \infty} \frac{1}{T} \int_0^T \theta_k' \, dt$$

over some trajectory of (4), provided that (12) converges. Note that this definition yields θ' if θ' is constant and is, a-priori, dependent on the trajectory. To compute (12) requires going to the full Equations (4). However, to lowest order, the frequency difference

$$(13) \qquad \lim_{T \to \infty} \frac{\varepsilon}{T} \int_0^T \dot{\phi}_k \, d\tau$$

can be computed from trajectories of (5). For any trajectory in the basin of attraction of the limit cycle of (10), (13) reduces to

$$(14) \qquad \frac{\varepsilon}{T_0} \int_0^{T_0} \dot{\phi}_k \, d\tau$$

where $T_0 = T_0(\beta)$ is the period of the limit cycle, and the integration of $\dot{\phi}_k$ is done along the limit cycle. Now

$$(15) \qquad \int_0^{T_0} \dot{\phi}_k \, d\tau = \tilde{\phi}_k(T_0) - \tilde{\phi}_k(0)$$

where $\tilde{\phi}_k$ denotes the "lift" of ϕ_k , i.e., the values obtained by inte-grating $\dot{\phi}_k$ without identifying mod 2π. Since the cycle is homotopic to the circle $\phi_k = 0$ $k \neq j$, $\tilde{\phi}_k(T_0) = \tilde{\phi}_k(0)$ $\forall k \neq j$. This means that, to $O(\varepsilon^2)$ in the original time variable, all the oscillators $\theta_1, \ldots, \theta_j$ have the same frequency, and so do $\theta_{j+1}, \ldots, \theta_n$. The frequency jump is com-puted from (14) and (15) to be $2\pi\varepsilon/T_0(\beta)$, which is $O(\varepsilon)$ in the original

time t. Note, that, as $\beta \to \beta_0^+$, $T_0(\beta) \to \infty$, so for $\beta \approx \beta_0$, the frequency jump increases from zero as β increases. In particular, to low-
est order, there is not necessarily a rational relationship between the
frequencies on the two plateaus.

The following questions remain open for this simplest
($H = \sin\phi$, $\alpha = 1$) model:

1. What happens when β gets still bigger? Numerical calculations
suggest that more plateaus form. (See Fig. 4.) Does this correspond to
the existence of invariant T^i T^n for (9), where i+1 is the number of
plateaus? If so, what is the dynamics on the invariant subtorii?

2. Work out the details when n is even. In that case, the two
middle components of ξ_0 reach $-\pi/2$ simultaneously, and critical points
coalesce in foursomes instead of pairs.

3. Suppose the gradient in underlying frequencies is not linear.
How does this affect the behavior?

It should be pointed out that for a linear gradient in natural
frequency and $H = \sin\phi$ (or any nearby odd function of ϕ), the frequency
plateaus are constrained by symmetry consideration to be symmetric with
respect to the midpoint of the gradient (e.g. the phase-locked solution
always has as its frequency the average of the natural frequencies).
However, in the intestine data, the plateaus lie at or above the natural
frequency. We believe that this effect can be accounted for by a phase
model, but one must have more complicated oscillators and/or coupling;
anisotropy in the coupling also changes the qualitative picture.

If we stick with $H = \sin\phi$, it is relatively easy to see some of
the effects of anisotropy: Equation (5) becomes

(16) $$\dot{\phi}_k = -\beta + \sin\phi_{k+1} - (\alpha+1)\sin\phi_k + \alpha\sin\phi_{k-1}$$

with $H(-\phi_0) = 0 = H(\phi_{n+1})$. It can be shown [11] that when β is small
enough that (16) has a stable critical point ξ_0 , the associated frequency of this phase-locked state depends on α: for $\alpha > 1$, the forward
coupling is stronger, and the frequency is closer to that of the initial,
i.e., high, end. (It is, however, less than the highest frequency.)
Similarly, $\alpha < 1$ skews the frequency to the lower end. (See Fig. 5.)
The graph of the components ξ_{0k} of ξ_0 vs. k is also skewed: e.g. for
$\alpha > 1$, $\max|\xi_{0k}|$ occurs for $k > j \equiv (n+1)/2$, which suggests that, for
larger β, the break between the frequency plateaus will occur in the
lower frequency range. (See Fig. 6.)

It is considerably harder to understand what is happening when the
coupling $H(\phi)$ in the phase model is more complicated, in particular no

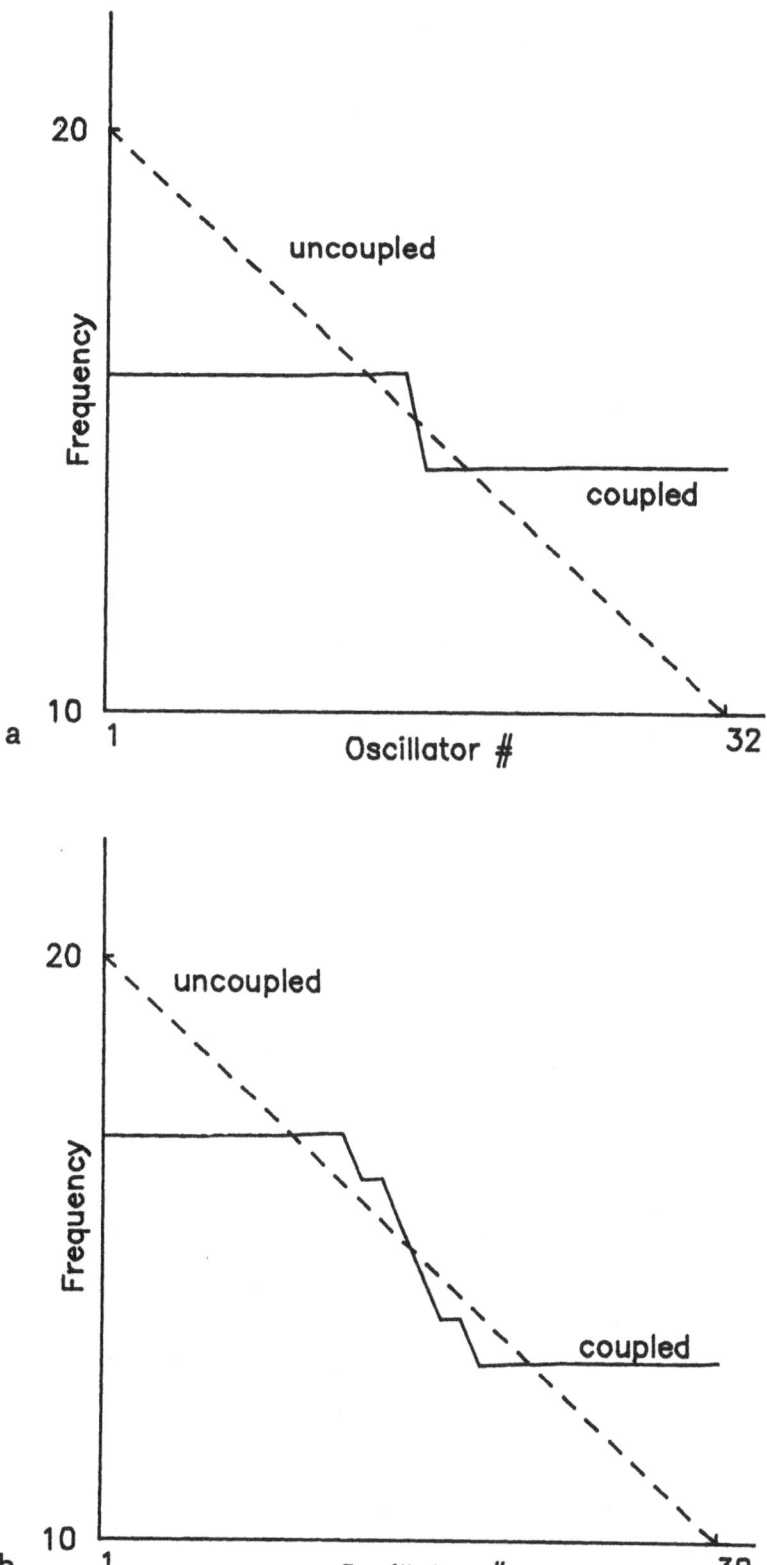

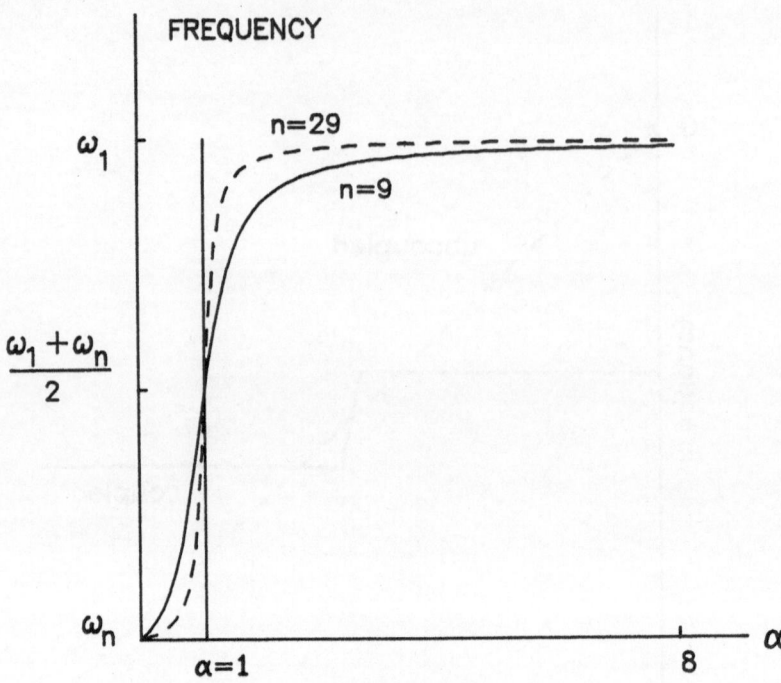

Figure 5
The frequency of the phase-locked solution as a function of the amount
of the amount of anisotropy, n=9 and n=29.

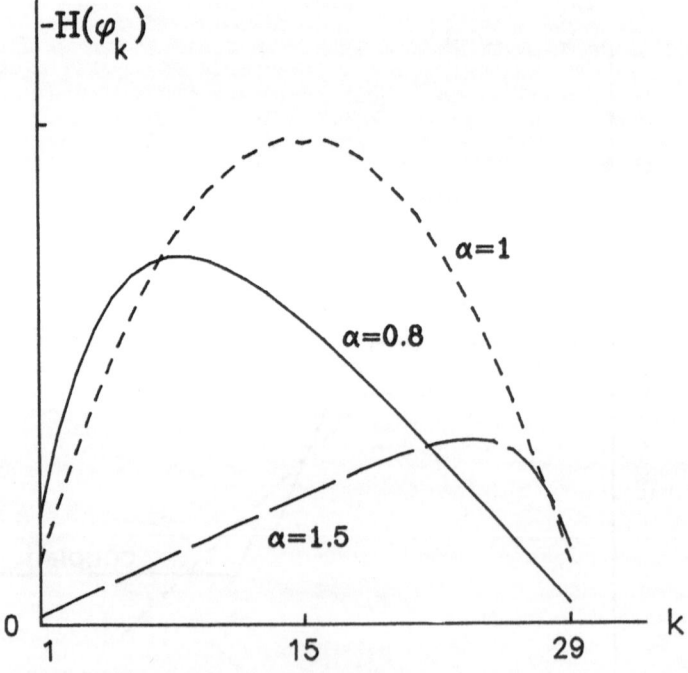

Figure 6
The graphs of $-\sin\phi_k^-$ vs. k for various α , n=29.

longer an odd function of ϕ (for example, the coupling (7) associated with the oscillation (6), B $\neq$ 0). It is no longer easy even to calculate the phase-locked solution. Paradoxically, however, it may be easier to get some results about this case when n is large than when n is small. For consider the following version of (5) when α = 1, where H(ϕ) is written as $H_e(\phi) + H_0(\phi)$ (i.e., a decomposition of H into its even and odd part):

(17)
$$\dot{\phi}_k = \Delta_k + H_e(\phi_{k+1}) - H_e(\phi_{k-1})$$
$$+ H_0(\phi_{k+1}) - 2H_0(\phi_k) + H_0(\phi_{k-1}).$$

If $\Delta_k = -\beta/n$ and n is large, the R.H.S. of (17) may be written approximately as an equation in a continuous variable:

(18)
$$\phi_t(x) = \frac{1}{n}\{-\beta + [H_e(\phi)]_x + \frac{1}{n}[H_0(\phi)]_{xx}\}$$

where $\phi_k = \phi(k/n)$, $0 \le x \le 1$, H($-\phi$) = 0 at x = 0 , H(ϕ) = 0 at x = 1. This suggests that a stable phase-locked solution to (17) should be close (e.g. in an L^1 norm) to a time-independent solution to (18), and numerical simulations corroborate this when $H_e(\phi) \not\equiv 0$. More details will be found in [17].

Note that the scaling $\Delta_k = -\beta/n$ of the frequency differences is such that the difference in frequency from the top to the bottom of the chain remains constant as n $\to \infty$. Thus if the formal analysis is accurate, for a fixed (not too large) frequency difference, (17) remains phase-locked as n $\to \infty$. This contrasts with H = sinϕ, for which the total difference in frequencies at β_0 (when phase-locking is lost) is O(1/n); for any fixed β, the phase-locked solution disappears for n sufficiently large.

We conjecture that the dynamics of (5) for $H_e(\phi) \not\equiv 0$ is also quite different, and may contain horseshoes. Relevant to this is an example of Keener and Hoppensteadt of a phase model involving 3 oscillators, and a coupling function less symmetrical than sinϕ. This example has a stable critical point (phase-locked solution), but the set of points attracted to this point omits, not a union of manifolds, but a Cantor set [18].

References

1. Connor, J., "On exploring the basis for slow potential oscillations in the mammalian stomach and intestine," J. Exp. Biol. 81, 153-173 (1979).

2. Daniel, E. and Sarna, S., "The generation and conduction of activity in smooth muscle," Ann. Rev. Pharmacol. Toxicol. 18, 145-66 (1978).

3. Connor, J., Mangel, A. and Nelson, B., "Propagation and entrainment of slow waves in cat small intestine," Amer. J. Physiol. 237, C237-C246 (1979).

4. Diamont, N. and Bortoff, "Effects of transection on the intestinal slow-wave frequency gradient." Am. J. Phys. 216, 734-743 (1969).

5. Diamond, N.E., Rose, P.K. and Davison, E.J., "Computer simulation of intestinal slow-wave frequency gradient," Amer. J. of Physiology 219, 1684-1690 (1970).

6. Sarna, S.K., Daniel, E.E. and Kingma, Y.J., "Simulation of slow-wave electrical activity of small intestine," Amer. J. of Physiology 221, 166-175 (1971).

7. Robertson-Dunn, B. and Linkens, D.A., "A mathematical model of the slow-wave electrical activity of the human small intestine," Medical and Biological Engineering, 750-757 (Nov. 1974).

8. Brown, B.H., Duthie, H.L., Horn, A.R. and Smallwood, R.H., "A linked oscillator model of electrical activity of human small intestine," Amer. J. Physiology 229, 384-388 (1975).

9. Patton, R.J. and Linkens, D.A., "Hodgkin-Huxley type electronic modelling of gastrointestinal electrical activity," Med. & Biol. Eng. & Computing, 16, 195-202 (1978).

10. Keener, J., lecture, Univ. of Utah (1976).

11. Winfree, A.T., The Geometry of Biological Time, Springer Verlag, NY (1980).

12. Ermentrout, G.B. and Kopell, N., "Frequency plateaus in a chain of weakly coupled oscillators, I," to appear.

13. Fenichel, N., "Persistence and smoothness of invariant manifolds for flows," Indiana U. Math. J. 21, 193-226 (1971).

14. Hirsch, M.W., Pugh, C.C. and Shub, M., Invariant Manifolds, Lecture Notes in Mathematics 583, Springer-Verlag, NY (1977).

15. Hale, J., Ordinary Differential Equations, John Wiley, NY (1969).

16. Kopell, N. and Howard, L.N., "Bifurcations and trajectories joining critical points," Advances in Math., 18, 306-358 (1975).

17. Ermentrout, G.B. and Kopell, N., "Frequency plateaus in a chain of coupled oscillators, II," to appear.

18. Keener, J., pers. comm.

Numerical Hopf Bifurcation Computation

for

Delay Differential Systems

B. Hassard

Dept. of Mathematics

SUNY at Buffalo

Abstract

Two schemes for the analysis of Hopf bifurcation in delay differential systems are described. The first scheme approximates the delay differential systems by ordinary differential systems, then uses the code BIFOR2 to analyze the ordinary differential systems. The second scheme analyzes delay differential systems directly and has been implemented in a new code, BIFDD.

1. Introduction

The theme of the talk associated with this paper was, how to apply an existing code for the analysis of Hopf bifurcation in ordinary differential systems, to integro-differential and delay differential systems for which the code was not originally designed. The interest in integro-differential systems was largely restricted to the use of such systems in approximating delay differential systems. A conclusion of the talk was, that although the existing code could be successfully applied, it would be more convenient and efficient to apply a code especially designed for such systems. Since the talk, a code for the analysis of Hopf bifurcation in delay differential systems has been written. This paper includes both the material of the talk, and a description of the new code.

2. <u>Hopf bifurcation for ordinary differential systems</u>

To introduce Hopf bifurcation , we shall briefly review the ordinary differential case.

Consider a system of ODE's

$$\dot{x} = f(x;v)$$

where x is in R^n , v is a real parameter, and f is at least C^3 jointly in x, v . The system is assumed to have a stationary solution $x_*(v)$ (a solution of $f(x;v) = 0$) for v in some interval. Let A(v) denote the Jacobian matrix of f with respect to x at the stationary solution, and let $\lambda_1(v)$, $\lambda_2(v)$,..., $\lambda_n(v)$ denote the eigenvalues of A(v) .

Suppose there is a value $v = v_c$ (subscript c for critical) such that

$$\lambda_1(v) = \bar{\lambda}_2(v) = \alpha(v) + i\,\omega(v) ,$$

$$v \text{ near } v_c ,$$

$$\alpha(v_c) = 0 , \alpha'(v_c) \neq 0$$

$$\omega(v_c) > 0 , \text{ and}$$

$$\text{Re } \lambda_j(v_c) < 0 \text{ for } j = 3,..,n.$$

Then, by Hopf's theorem [1,2] there is a family of periodic solutions

$$x = p_\varepsilon(t)$$

$$= x_*(\nu_c) + \varepsilon \text{ Re}\{ v_1 e^{2\pi i t/T(\varepsilon)} \} + O(\varepsilon^2)$$

The solution $p_\varepsilon(t)$ occurs for

$$\nu = \nu_c + \mu(\varepsilon)$$

$$= \nu_c + \mu_2 \varepsilon^2 + O(\varepsilon^4) \quad ,$$

and is $T(\varepsilon)$-periodic, where the period $T(\varepsilon)$ expands as

$$T(\varepsilon) = (2\pi/\omega(\nu_c))[1 + \tau_2{}^2 + O(\varepsilon^4)]$$

One of the characteristic exponents $\beta(\varepsilon)$ associated with $p_\varepsilon(t)$ expands as

$$\beta(\varepsilon) = \beta_2 \varepsilon^2 + O(\varepsilon^4) \quad ,$$

and if $\beta(\varepsilon) < 0$, $p_\varepsilon(t)$ is orbitally asymptotically stable with asymptotic phase.

3. The code BIFOR2

In a Hopf bifurcation calculation for ordinary differential systems along the lines of Ch. 3 of [2] the objectives are 1) to locate the critical value v_c and 2) to compute μ_2 , τ_2 and β_2 . The number μ_2 , if nonzero, determines the direction of bifurcation (+1 if $\mu_2 > 0$, -1 if $\mu_2 < 0$). The number β_2 , if nonzero, determines the stability of the family of periodic solutions $p_\varepsilon(t)$, for all sufficiently small ε . For relatively simple ordinary differential systems, it is possible to perform Hopf bifurcation calculations by hand: see the 'recipe-summary' in Ch. 2 of [2] . For more complicated systems, hand calculations tend to become impractical, either because of inability to obtain closed form expressions or because the volume of computation is overwhelming.

There is a code available, called BIFOR2, which will automatically perform Hopf bifurcation computations, both location of v_c and determination of μ_2, τ_2, β_2 (and $x_*(v_c)$, v_1, $\omega_0 = \omega(v_c)$, $\alpha'(v_c)$ and $\omega'(v_c)$ as well).

BIFOR2 is documented in [2] , Ch. 3 and the FORTRAN source is given on microfiche in [2] .

To apply BIFOR2, the user writes a subroutine which defines the ordinary differential system. The user also writes a main program which defines initial estimates of v_c and $x_*(v_c)$, then calls BIFOR2. BIFOR2 does the rest.

Although BIFOR2 was designed for ordinary differential systems, it has also been applied to partial-differential, integro-differential and delay differential systems, [2,Chs. 3,4,5; also 3]. The basic idea is to approximate the given system by an ordinary differential system, then analyze the approximating system with BIFOR2.

4. <u>Approximation of delay differential systems by ordinary differential systems</u>

Consider the delay differential system

$$\dot{x}(t) = f(x(t-t_1),..,x(t-t_M);v)$$

where x is in R^n , $t_j \geq 0$, j=1,..,M are fixed 'delays' (t_1 might = 0) and v is real, the bifurcation parameter. One way to analyze Hopf bifurcation in such a system is to 1) approximate the delay differential system by an integro-differential system, 2) reduce the integro-differential system to an ordinary differential system, then 3) analyze the ordinary differential system with BIFOR2. To accomplish 1), each term $x(t-t_j)$ where $t_j > 0$ is approximated by an integral $\int_{-\infty}^{t} K_{n,j}(t-s) \, x(s) \, ds$, where $K_{n,j}(t)$ satisfies a homogeneous, ordinary differential equation with constant coefficients.

For example, if $t_j = 1$, then

$$x(t-1) \equiv c_n \int_{-\infty}^{t} (t-s)^n \, e^{-n(t-s)} \, x(s) \, ds,$$

where $c_n = n^n / n!$. (Exercise: if $x(t)$ is continuous and
bounded, then show that the l.h.s. is the limit of the r.h.s.)
The kernel

$$K_n(t) = n^n \, t^n \, e^{-nt} \, /n!$$

satisfies

$$(d/dt + n)^{n+1} \, K_n(t) = 0 \, ,$$

so any delay differential system

$$\dot{x} = f(x(t), x(t-1); \nu)$$

may be approximated by

$$\dot{x} = f(x(t), \int_{-\infty}^{t} K_n(t-s) \, x(s) \, ds; \nu)$$

which is equivalent by the 'trick' described below to an ordinary
differential system, and so may be analyzed with BIFOR2. The
'trick' is based upon the observation that any term
$\int_{-\infty}^{t} K(t-s) \, x(s) \, ds$ in an integro–differential system, where $K(t)$
satisfies a homogeneous ordinary differential equation with
constant coefficients, may be eliminated. Of course, a price must
be paid, here the introduction of additional variables. Suppose
that $K(t)$ satisfies the ordinary differential equation

$$K^{(p+1)}(t) = \sum_{0}^{p} a_k \, K^{(k)}(t), \quad a_k\text{'s constants.}$$

Let $y^j(t) = \int_{-\infty}^{t} K^{(j)}(t-s) \, x(s) \, ds, \; j=0,..,p$. Then the term
$\int_{-\infty}^{t} K(t-s) \, x(s) \, ds$ in the integro-differential system is replaced

with $y^0(t)$, and the differential equations

$$\dot{y}^j = y^{j+1} + K^{(j)}(0) \ x(t), \quad j=0,..,p-1$$

$$\dot{y}^p = \sum_0^p a_k \ y^k + K^{(p)}(0) \ x(t)$$

are added to the system. (The basic trick is quite old, but the application in conjunction with extrapolation to approximate delta function kernels is apparently new.)

5. Example of the approximation procedure: a system with two delays

To illustrate this 3 step procedure for analysis of Hopf bifurcation in a delay differential system, we shall sketch the application to the system

$$dx/dt = -\nu [x(t-1) + x(t-2)](1 - x^2(t)).$$

studied by Jones [4], and Kaplan and Yorke [5] ; see also [2], Ch 4. First, we choose an integer n 'large'. Let

$$y_j(t) = \int_{-\infty}^t K_n^{(j)}(t-s) \ x(s) \ ds, \quad j = 0,..,n$$

where $K_n(t)$ is as above. Then $y_0(t) \cong x(t-1)$. To approximate $x(t-2)$, let $L_n(t) = K_n(t/2)/2$ and let

$$z_j(t) = \int_{-\infty}^t L_n^{(j)}(t-s) \ x(s) \ ds, \quad j = 0,..,n.$$

Then, $z_0(t) \cong x(t-2)$. The approximate ordinary differential system is then,

$$dx/dt = - (y_0 + z_0)(1 - x^2)$$

$$dy_j/dt = y_{j+1} , \quad j = 0,..,n-1$$

$$dy_n/dt = - \sum_0^n C_{n+1,k} \, n^{n+1-k} \, y_k(t) + K_n^{(n)}(0) \, x(t)$$

$$dz_j/dt = z_{j+1} , \quad j = 0,..,n-1$$

$$dz_n/dt = - \sum_0^n C_{n+1,k} \, (n/2)^{n+1-k} \, z_k(t) + L_n^{(n)}(0) \, x(t)$$

This system is of the form $\dot{X} = F(X;\nu)$ where X is the $N = 2n+3$ dimensional vector

$$X = (x,y_0,..,y_n,z_0,..,z_n).$$

A program was written to apply BIFOR2 to this system. The subroutine required to define $F(X;\nu)$ and the Jacobian of F with respect to X, was 60 statements long. The computation was performed for $n = 1,..,8$, so the ordinary differential systems analyzed were of order $5,7,..,19$. The following table of approximate results $\nu_c^{(n)}$, $\mu_2^{(n)}$, $\beta_2^{(n)}$ were obtained. ($\tau_2^{(n)}$ was always 0.)

n	$\nu_c^{(n)}$	$\mu_2^{(n)}$	$\beta_2^{(n)}$
1	.972002	.243000	-.075052
2	.830264	.207566	-.103413
3	.764451	.191113	-.118192
4	.726407	.181602	-.128790
5	.702277	.175569	-.137028
6	.685848	.171462	-.143599
7	.674026	.168507	-.148935
8	.665148	.166287	-.153339

The convergence is rather slow. To speed up the convergence, Richardson extrapolation [6] was performed. In the following tables, column j+1 was obtained by Richardson extrapolation of column j, j=1,2,3,4. The assumption upon which the extrapolation is based is that the truncation error expands in inverse powers of n. The truncation error in column j is then $O(1/n^j)$. The most accurate approximations in each table are the bottom right entries. The values for ν_c , μ_2 and β_2 computed analytically are (rounded to 6 decimal places) ν_c = .604600, μ_2 = .151150, β_2 = -.197548. The numerical results obtained from the 3 step process thus confirm the analytical results.

Richardson extrapolation of $\nu_c^{(n)}$

.972002	.688526	.614260	.593196	.598359
.830264	.632826	.595829	.598036	.602962
.764451	.612272	.597382	.601989	.604205
.726407	.605758	.600046	.603504	.604486
.702277	.603701	.601733	.604084	
.685848	.603100	.602224		
.674026	.603000			
.665148				

Richardson extrapolation of $\nu_2^{(n)}$

.243000	.172131	.153565	.148299	.149590
.207566	.158207	.148957	.149509	.150741
.191113	.153068	.149346	.150497	.151051
.181602	.151439	.150011	.150876	.151121
.175569	.150925	.150433	.151021	
.171462	.150774	.150681		
.168507	.150750			
.166287				

Richardson extrapolation of $-\beta_2^{(n)}$

.075052	.131774	.153075	.173389	.187688
.103413	.147749	.170850	.186795	.193576
.118192	.160583	.182070	.192236	.195637
.128790	.169984	.187948	.194561	.196521
.137028	.176451	.191175	.195719	
.143599	.180950	.193089		
.148935	.184170			
.153339				

6. Discussion of the approximation procedure

The three step procedure just described was developed in order to independently confirm the exact, analytical calculations of [2], Ch. 4. Since the systems analyzed involved at most 2 time delays, the order of the approximating ordinary differential equations was within the capabilities of the code BIFOR2. For systems with more time delays, however, the limits of BIFOR2 will be reached. Also, the procedure is clumsy because extrapolation is needed to obtain adequately accurate results, and because the user must perform the analytic work to recast the delay differential system as an approximating ordinary differential system. At the time [2] was written, it was clear that a code specifically designed for Hopf bifurcation analysis of delay

differential systems could be designed along the lines of BIFOR2.
It was also clear that the code would be substantially different
from BIFOR2. The algebraic eigenvalue problem which arises in
delay differential systems is nonstandard, so EISPACK routines are
no longer applicable. A new scheme for locating the critical
value of the bifurcation parameter had to be developed. Instead
of a single Jacobian matrix as in the ordinary differential
situation, there is a set of Jacobian matrices, one for each
distinct lagged variable. The relevant inner product itself
involves these matrices, so different normalization and projection
schemes had to be developed. One idea that remains valid is that
of computing the required second and third order partial
derivatives by numerical differencing of the first order partial
derivatives as evaluated by the user-supplied subroutine.

7. The code BIFDD

Consider the delay differential system

$$\dot{x}(t) = f(x(t-t_1),..,x(t-t_M);\nu)$$

where x is in R^n , $t_j \geq 0$, $j=1,..,M$ are fixed 'delays'
(t_1 might $= 0$) and ν in R^1 is the bifurcation parameter.

A code, BIFDD, has now been written to analyze Hopf
bifurcations in such delay differential systems directly. The
user writes a subroutine which defines the delay differential
system. This subroutine must evaluate the function f and the set

of Jacobian matrices of f with respect to the variables $x_1 = x(t-t_1),..,$ $x_M = x(t-t_M)$. The user also writes a main program which defines initial estimates of ν_c, ω_0 and $x_*(\nu_c)$ and calls BIFDD.

BIFDD performs two main computations; 1) determines ν_c and $\omega_0 = Im\{\lambda_1(\nu_c)\}$, and 2) computes μ_2 , τ_2 and β_2 .

8. <u>Computation of</u> (ν_c ,ω_0) <u>within BIFDD</u>

To find the pair (ν_c,ω_0), BIFDD solves the pair of nonlinear algebraic equations

$$Re \quad det(\nu,\omega) \quad = \quad 0$$

$$Im \quad det(\nu,\omega) \quad = \quad 0$$

where det is the determinant of the matrix

$$i\omega I - \sum_{j=1}^{M} A_j(\nu) \ e^{-i\omega t_j}$$

The matrices $A_j(\nu)$, $j=1,..,M$ are the Jacobian matrices of $f(x_1,..,x_M;\nu)$ with respect to the vectors x_j, $j=1,..,M$ evaluated at $x_1=..=x_M= x_*(\nu)$.

The technique presently used to solve the pair of equations for (ν_c,ω_0) is a Newton iteration with Jacobians evaluated by one-sided numerical differencing.

At each different value of ν during the iteration for (ν_c, ω_0), the stationary point $x_*(\nu)$ is computed by solving

$$f(x:\nu) = 0$$

where $f(x;\nu) = f(x,x,..,x;\nu)$. Newtons method is used. The Jacobian matrix of $f(x;\nu)$ with respect to x is computed as

$$\sum_{j=1}^{M} A_j(x;\nu)$$

where for $j=1,..,M$, $A_j(x;\nu)$ is the Jacobian of $f(x_1,..,x_M;\nu)$ with respect to x_j, evaluated at $x_1=x_1=..=x_M= x$.

The user-supplied subroutine has the job of evaluating both $f(x_1,..,x_M;\nu)$ and the set of matrices $A_j(x_1,..,x_M;\nu)$, $j=1,..,M$, where $A_j(x_1,..,x_M;\nu)$ is the Jacobian of $f(x_1,..,x_M;\nu)$ with respect to x_j. In terms of $A_j(x_1,..,x_M;\nu)$, $A_j(x;\nu) = A_j(x,..,x;\nu)$. In terms of $A_j(x;\nu)$, $A_j(\nu) = A_j(x_*(\nu);\nu)$.

9. Computation of μ_2, τ_2 and β_2.

9.1 Introduction

We refer the reader to [2 , Ch. 4, Section 2] for the theory and the general algorithm used in computing μ_2, τ_2 and β_2. The algorithm described below is specialized to delay differential systems, and has been simplified. The structure of the computation parallels that described in [2 , Ch. 3] for ordinary differential systems.

To determine β_2 , it is necessary to know $c_1(0)$, the coefficient of the cubic term in the Poincare normal form for the 2nd order ordinary differential equation obtained by restricting the delay differential system to the slice $\nu = \nu_c$ of the center manifold for the delay differential system [2]. To determine μ_2 and τ_2 , it is necessary to know, in addition, $\alpha'(\nu_c)$ and $\omega'(\nu_c)$, the real and imaginary parts of $\lambda_1'(\nu_c)$.

9.2 <u>Computation of</u> $\lambda_1'(\nu_c)$

The eigenvalue $\lambda_1(\nu)$ satisfies

$$\det(\lambda_1;\nu) = 0$$

where now det is the determinant of the matrix

$$\lambda_1 I - \sum_{j=1}^{M} A_j(\nu) \ e^{-\lambda_1 t_j}.$$

The derivative, $\lambda_1'(\nu_c)$ is approximated by the symmetric difference quotient

$$(\lambda_1(\nu_c+h) - \lambda_1(\nu_c-h))/2h$$

in which each of $\lambda_1(\nu_c+h)$ and $\lambda_1(\nu_c-h)$ are found by solving the pair of real nonlinear algebraic equations equivalent to the single complex equation $\det(\lambda_1;\nu) = 0$. The same technique (in fact, the same code) is used to solve for α and ω (given ν) as is used to solve for (ν_c,ω_0).

9.3 Computation of $q(\theta)$ and $q^*(s)$

The linearization of the delay differential system about the stationary point for the critical value of ν has the form

$$\dot{y} = \sum_{j=1}^{M} A_j(\nu_c) \ y(t-t_j).$$

The solution of this system corresponding to the eigenvalue $i\omega_0$ is $y(t) = q(t)$, where

$$q(\theta) = v_1 \ e^{i\omega_0\theta}$$

and v_1 is the right null vector of the matrix

$$i\omega_0 I - \sum_{j=1}^{M} A_j(\omega_c) \ e^{-i\omega_0 t_j}$$

The solution $q^*(s)$ of the adjoint system, used for projection in the direction of $q(\theta)$ is

$$q^*(s) = \bar{u}_1 \ e^{i\omega_0 s}$$

where u_1 is the left null vector of the same matrix.

Vectors v_1 and u_1 are each computed by inverse iteration. Only one LU factorization is needed to compute both vectors. During the inverse iterations, v_1 and u_1 are each normalized to have Euclidean norm one. Once the inverse iterations are done, v_1 is renormalized such that the first nonzero element is one. Also, u_1 is renormalized relative to v_1 so that the inner product $<q^*,q> = 1$, where

$$<q^*, q> = u_1 \cdot v_1 + \sum_{j=1}^{M} t_j e^{-i\omega_0 t_j} u_1 A_j(v_c) v_1$$

9.4 Computation of f_{20}, f_{11}, f_{02}, g_{20}, g_{11} and g_{02}

Let $f_0(z, \bar{z})$ denote the restriction of f to the slice $v = v_c$ of the center manifold; that is, let

$$f_0(z, \bar{z}) = f(x_1(z, \bar{z}), \ldots, x_M(z, \bar{z}) ; v_c)$$

where

$$x_j(z, \bar{z}) = x_* + z \, q(t_j) + \bar{z} \, \bar{q} \, (-t_j) + w(z, \bar{z}, -t_j)$$

and

$$w(z, \bar{z}, \theta) = w_{20}(\theta) z^2/2 + w_{11}(\theta) z \bar{z} + w_{20}(\theta) \bar{z}^2/2$$
$$+ O(|z|^3)$$

is the function describing the slice. (The coefficients in the expansion of w are not yet known.) The vectors f_{20}, f_{11} and f_{02} are the second order partial derivatives of f_0, at $z = \bar{z} = 0$, that is

$$f_{20} = \partial^2 f_0/\partial z^2,$$

$$f_{11} = \partial^2 f_0/\partial z \, \partial \bar{z},$$

$$f_{02} = \partial^2 f_0/\partial \bar{z}^2.$$

Since $w(z, \bar{z}, \theta) = O(|z|^2)$, the second order partials of f_0 at the origin are the same as those of

$$\mathbf{f}_0(z, \bar{z}) = f(\tilde{x}_1(z, \bar{z}) ,..,\tilde{x}_M(z, \bar{z}) ; \nu_c)$$

where

$$\tilde{x}_j(z, \bar{z}) = x_* + z \, q(-t_j) + \bar{z} \, \bar{q} \, (-t_j).$$

Now,

$$\partial \mathbf{f}_0/\partial z \, (z, \bar{z}) = \sum_{j=1}^{M} A_j(\tilde{x}_1(z, \bar{z}) ,...,\tilde{x}_M(z, \bar{z}) ; \nu) q(-t_j)$$

where the matrices A_j are evaluated by calling the user-supplied subroutine, and $q(\theta)$ is as computed above. The vectors f_{20} and f_{11} are then computed in 2 steps: 1) symmetric differencing with respect to $y_1 = \mathrm{Re}\{z\}$ and $y_2 = \mathrm{Im}\{z\}$ to approximate $(\partial/\partial y_1)(\partial f_0/\partial z)$ and $(\partial/\partial y_2)(\partial f_0/\partial z)$, 2) use of the symbolic formulae

$$\partial/\partial z = (\partial/\partial y_1 - i \, \partial/\partial y_2)/2 ,$$

$$\partial/\partial \bar{z} = (\partial/\partial y_1 + i \, \partial/\partial y_2)/2$$

to form $(\partial/\partial z)(\partial f_0/\partial z)$ and $(\partial/\partial \bar{z})(\partial f_0/\partial z)$. Since f_0 is real, $f_{02} = \bar{f}_{20}$ and it is not necessary to compute f_{02} separately.

The complex numbers g_{20}, g_{11} and g_{02} are the second partial derivatives

$$g_{20} = \partial^2 g /\partial z^2$$

$$g_{11} = \partial^2 g /\partial z \, \partial \bar{z}$$

$$g_{02} = \partial^2 g / \partial \bar{z}^2 \qquad \text{at} \quad z = \bar{z} = 0,$$

where $g(z, \bar{z}) = u_1 \cdot f_0(z, \bar{z})$. These second partial derivatives are simply computed as

$$g_{20} = u_1 \cdot f_{20} ,$$

$$g_{11} = u_1 \cdot f_{11} ,$$

$$g_{02} = u_1 \cdot \bar{f}_{20} .$$

9.5 Computation of $w_{20}(\theta)$, $w_{11}(\theta)$ and $w_{02}(\theta)$

The coefficient functions $w_{20}(\theta)$ and $w_{11}(\theta)$ are of the form,

$$w_{20}(\theta) = w_{201} q(\theta) + w_{202} \bar{q}(\theta) + W_{20} e^{2i\omega_0\theta}$$

$$w_{11}(\theta) = w_{111} q(\theta) + w_{112} \bar{q}(\theta) + W_{11}$$

where w_{201}, w_{202}, w_{111} and w_{112} are complex numbers and W_{20} is a complex n-vector. Since the function $w(z, \bar{z}, \theta)$ is a real, n-vector valued function, the coefficient function $w_{11}(\theta)$ of $z \bar{z}$ in the Taylor expansion of w is also in R^n. This implies that $W_{112} = \bar{W}_{111}$ and W_{11} is in R^n. Since $w_{20}(\theta)$ and $w_{02}(\theta)$ are coefficients of $z^2/2$ and $\bar{z}^2/2$ (respectively) in the Taylor expansion of w, $w_{02}(\theta) = \bar{w}_{20}(\theta)$ and it is not necessary to compute w_{02} separately.

The numbers W_{201}, W_{202}, W_{111} and W_{112} are computed as

$$W_{201} = -(u_1 \cdot f_{20})/i\omega_0$$

$$W_{202} = -(\bar{u}_1 \cdot f_{20})/3i\omega_0$$

$$W_{111} = (u_1 \cdot f_{11})/i\omega_0$$

$$W_{112} = \bar{W}_{111} .$$

The vector W_{20} is computed by solving the complex linear system

$$[2i\omega_0 I - \sum_{j=1}^{M} A_j(v_c) e^{-2i\omega_0 t_j}] W_{20} = f_{20}$$

and the vector W_{11} is computed by solving the real linear system

$$[- \sum_{j=1}^{M} A_j(v_c)] W_{11} = f_{11} .$$

9.6 Computation of g_{21}

The complex number g_{21} is the third partial derivative

$$g_{21} = \partial^3 g/\partial z^2 \partial \bar{z} \quad \text{at} \quad z = \bar{z} = 0 ,$$

where

$$g(z, \bar{z}) = u_1 \cdot f_0(z, \bar{z}) .$$

The third order partials of g at the origin are the same as those of

$$\hat{g}(z,\ \bar{z})\quad =\quad u_1.f(\hat{x}_1(z,\ \bar{z})\ ,..,\ \hat{x}_M(z,\ \bar{z})\ ;\ \nu_c)$$

where

$$\hat{x}_j(z,\ \bar{z})\quad =\quad x_* \ +\ z\ q(-t_j)\ +\ \bar{z}\ \bar{q}\ (-t_j)\ +$$

$$+\ (z^2/2)\ w_{20}(-t_j)\ +\ z\ \bar{z}\ w_{11}(-t_j)\ +\ (\ \bar{z}^2/2)\ w_{02}(-t_j)$$

Now,

$$\partial\hat{g}/\partial z\ (z,\ \bar{z})\quad =\quad \sum_{j=1}^{M}\ u_1\ A_j(\hat{x}_1(z,\ \bar{z})\ ,..,\ \hat{x}_M(z,\ \bar{z})\ ;\ \nu_c)$$

$$[\ q(-t_j)\ +\ z\ w_{20}(-t_j)\ +\ \bar{z}\ w_{11}(-t_j)\ ]\ .$$

Symbolically,

$$(\ \partial/\partial z\ )(\ \partial/\partial\ \bar{z}\ )\quad =\quad (1/4)(\ \partial^2/\ \partial y_1^2\ +\ \partial^2/\ \partial y_2^2)$$

where y_1 = Re$\{z\}$ and y_2 = Im$\{z\}$. To compute g_{21} , numerical
differencing is used to approximate the Laplacian of $\partial\hat{g}/\partial z$ with
respect to y_1 and y_2 . The user-supplied subroutine is called to
evaluate the matrices A_j , and $\partial\hat{g}/\partial z$ is computed according to the
formula above. The standard 9-point difference approximation to
the Laplacian is employed.

9.7 Evaluation of $c_1(0)$, μ_2, τ_2 and β_2.

These coefficients are evaluated by the usual formulae

$$c_1(0) \;=\; (i/2\omega_0)(g_{20}g_{11} - 2|g_{11}|^2 - |g_{02}|^2/3) + g_{21}/2 \;,$$

$$\mu_2 \;=\; -(\text{Re } c_1(0))/\alpha'(\nu_c)$$

$$\tau_2 \;=\; -(\text{Im } c_1(0) \;+\; \mu_2\, \omega'(\nu_c)\,)/\omega(\nu_c) \;.$$

$$\beta_2 \;=\; 2 \text{ Re } c_1(0)$$

9.8 The form of the periodic solutions

The family of bifurcating periodic solutions of the delay differential system is then of the form,

$$p_e(t) \;=\; x_*(\nu_c) \;+\; 2\varepsilon\text{Re}\{v_1 e^{i\omega_0 t}\} \;+\; 0(\varepsilon^2)$$

where ε and ν are related by

$$\nu \;=\; \nu_c + \mu_2\,\varepsilon^2 \;+\; 0(\varepsilon^4).$$

The period $T(\varepsilon)$ and the characteristic exponent of interest are $\beta(\varepsilon)$ are of the form,

$$T(\varepsilon) \;=\; (2\pi/\omega_0)(1 \;+\; \tau_2\varepsilon^2 \;+\; 0(\varepsilon^4)),$$

$$\beta(\varepsilon) \;=\; \beta_2\varepsilon^2 \;+\; 0(\varepsilon^4).$$

Note the 2ε instead of ε in the expression for the periodic solutions. This factor arises because of the definition of $q(\theta)$

in terms of v_1 above: the present definition is consistent with Ch. 4 of [2]. For the computation to reduce in the special case $M = 1$, $t_1 = 0$ to the analogous computation for ordinary differential systems in Ch. 3 of [2], it is necessary to take instead

$$q(\theta) = (v_1/2) \, e^{i\omega_0 \theta} \, .$$

10. Discussion

The code BIFOR2 has performed reliably for several years, and we have much confidence in the results obtained by the approximation technique described in the first part of this paper. The new code BIFDD represents a more elegant way of analyzing Hopf bifurcation in delay differential systems, but at present writing, has been tested only on the delay differential systems in Ch. 4 of [2]. The code should be able to handle with ease systems of order 10 or 20 with 5 or 10 distinct delays.

The code BIFDD can be applied to ordinary differential as well as delay differential systems. BIFDD is not, however, a replacement for BIFOR2. BIFDD requires the user to provide an estimate of $\omega(v_c)$, and computes only the eigenvalue $\lambda_1(v)$ during the location of v_c. BIFOR2 does not need the estimate of $\omega(v_c)$, and provides the option to compute all of the eigenvalues during the location of v_c. These differences occur because the relevant linear operator in the general delay differential computation may

have an infinite spectrum. The eigenvalue problem is a
nonstandard generalization of the algebraic eigenvalue problem,
and none of the usual algorithms (QR, LR, QZ, LZ) are applicable.
These differences also point up the potential of BIFDD to mislead:
BIFDD computes only one eigenvalue, so the Hopf hypotheses
[2,Ch. 4] on the rest of the spectrum might be violated without
the user being aware of the problem. This same uncertainty is
present in Hopf bifurcation computations for partial differential
systems [3]. A way to diminish the uncertainty is to compute more
of the spectrum, and this is one way in which the current version
of BIFDD could be improved.

There are codes available for the numerical integration of
delay differential systems, and for computing families of periodic
solutions along bifurcation curves. E. Doedel (Concordia U.,
Montreal) is expert in this area. The type of Hopf bifurcation
analysis described in the present paper complements, rather than
competes with, direct numerical computations of the families. The
computations described here provide local approximations (ε small)
to the families of periodic solutions which arise by Hopf
bifurcation. The techniques we have described are cheaper than
the direct techniques, and so have an advantage when it is desired
to see how varying the physical parameters in a system affects the
Hopf bifurcation. For information about the global behavior of
the families of periodic solutions, direct numerical computations
must be used.

Acknowledgement

Portions of this work were performed while the author was supported under NSF grants MCS 79-705790 and MCS 81-06657 .

References

[1] Marsden, J. E. and McCracken, M. 'The Hopf Bifurcation and Its Applications', Applied Math. Sciences, vol. 19, Springer-Verlag N. Y. (1976)

[2] Hassard, B. D., Kazarinoff, N. D. and Wan, Y-H, 'Theory and Applications of Hopf Bifurcation', London Math. Soc. Lecture Note Series 41, Cambridge U. Press (1981)

[3] Hassard, B. D. and El-Henawy, I. Numerical Hopf bifurcation analysis in nonlinear ordinary and partial differential systems from chemical reactor theory, Appl. Math and Computation 9 (1981) 75-92

[4] Jones, G. S. The existence of periodic solutions of $f'(x) = -af(x-1)(1+f(x))$, J. Math. Anal. and Applics. 5 (1962), 435-450

[5] Kaplan, J. L. and Yorke, J. A. Ordinary differential equations which yield periodic solutions of differential delay equations, J. Math. Anal. and Applics. 48 (1974), 317-324

[6] Dahlquist, G. and Bjorck, A. 'Numerical Methods', Prentice Hall, N.J. (1974)

Traveling Waves with Finitely Many Pulses in a Nerve Equation

John A. Feroe
Department of Mathematics
Vassar College
Poughkeepsie, New York 12601

Abstract

Some reaction-diffusion equations which model the nerve axon are
known to support a traveling wave with a solitary pulse. Conditions
which previously established the existence of solutions with two
pulses are here used to show the existence of solutions with n pulses
for any $n \geq 2$. Provided are extensive motivating geometric arguments
as well as the analytic results and explicitly computed solutions for
the piecewise linear FitzHugh-Nagumo equations.

1. Introduction

The expository nature of this discussion of a recent result concerning the existence of multiple pulse waves in nerve axon equations will preclude providing details of proofs, the complete justification of claims, or the precise meaning of such time honored phrases as "sufficiently small". Exploited in place of detail will be the opportunity to communicate the underlying motivation of the result, a luxury all too seldom afforded authors in print. A more formal and more general paper is in preparation for future publication.

In recent work, Evans, Fenichel and Feroe [6] identify conditions for the existence of a family of traveling wave solutions, solutions each of which consists of a pair of pulses. That result is extended here to an arbitrary number of pulses. In doing so ideas have been adapted from Feroe [9] (by way of Šil'nikov [20]) where infinite pulse trains are discussed. All of these results have as a conceptually unifying source the geometry of the stable and unstable manifolds for a stationary point of the relevant dynamical system.

The discussion begins in Section 2 with the definitions and development needed to state the principal result. Section 3 shows the plausibility of the result through primarily geometric and pictorial means. A comparable analytic discussion follows in Section 4. Both Sections 3 and 4 are divided into an initial explaination of the underlying structure followed by the demonstration of how interactions within that structure lead to the result. In conclusion, Section 5 contains the results of numerical calculations.

2. Definitions and the principal result

An idealized nerve axon will be treated here as an infinitely long cylindrical membrane. This uniform semipermeable membrane is assumed to separate concentrations of various ions. Of primary interest is the transmembrane voltage $v(\xi,\tau)$ measured at position ξ and time τ. The permeability of the membrane is reactive to changes in voltage and as a result the axon has the ability to support the propagation along its length of a characteristic local voltage disturbance called the action potential. Indeed, this ability is the axon's basic role as an element of the nervous system.

The general form of nerve axon equations is given [12] as

$$\frac{\partial^2 v}{\partial \xi^2} = \frac{\partial v}{\partial \tau} + f(v, \vec{w})$$
$$\frac{\partial \vec{w}}{\partial \tau} = \vec{g}(v, \vec{w}) \ .$$

(1)

The dimension and precise physical meaning of the vector $\vec{w}(\xi,\tau)$ varies from model to model. We assume a unique stable rest state which can be taken to be at $v = 0, \vec{w} = 0$ so that

$$f(0,0) = 0$$
$$\vec{g}(0,0) = 0 \ .$$

The two particular cases of these equations which have been of

primary interest to mathematicians and/or physiologists are as
follows:

1. The Hodgkin-Huxley equations [16], a system with $\dim \vec{w} = 3$
which models experimental data from the giant axon of the squid.
The components of $\vec{w}$ are related to the permeability of the
axon to sodium and potassium ions. The reader is referred to
[12] for the specific equations, their background and further
references.

2. The FitzHugh-Nagumo equations [11, 18], a system which takes
$\dim \vec{w} = 1$ yet still manages to model such crucial behavior of
the nerve as threshold effects and the propagation of wave forms.
Specifically these equations use

$$f(v,w) = v(1 - v)(a - v) + w \qquad 0 < a < 1$$
$$g(v,w) = bv - dw \qquad b > 0, \, d > 0$$

(2)

The cubic function f can be replaced by the piecewise linear

$$f(v,w) = -v + H(v - a) + w \qquad (3)$$

where $H(v - a)$ is the Heaviside step function

$$H(v - a) = \begin{cases} 1 & \text{if} \quad v \geq a \\ 0 & \text{if} \quad v < a \end{cases}$$

This modification was suggested by McKean [17] and because it
affords the ability to find explicit solutions it has proved to

be of great use in identifying and illustrating properties of other systems and of nerve equations in general [8, 9, 10, 19].

The formal analysis offered in Section 4 treats the piecewise linear version of the FitzHugh-Nagumo equations. The preliminary discussion is more general, and work in preparation extends the results formally to more general cases.

A bounded solution to (1) which represents a wave traveling with fixed form and constant velocity $c > 0$, that is, a traveling wave solution, satisfies the condition

$$(v(\xi,\tau), \vec{w}(\xi,\tau)) = (v(\xi + c\tau, 0), \vec{w}(\xi + c\tau, 0))$$

$$\overset{\text{def}}{=} (v_c(\eta), \vec{w}_c(\eta))$$

where $\xi + c\tau \overset{\text{def}}{=} \eta$. Since

$$\frac{\partial^2 v}{\partial \xi^2} = \tilde{v}_c \quad , \quad \frac{\partial v}{\partial \tau} = c v_c' \quad \text{and} \quad \frac{\partial \vec{w}}{\partial \tau} = c \vec{w}_c'$$

a traveling wave solution of (1) is equivalent to a bounded solution of the traveling wave equations

$$\tilde{v}_c = c v_c' + f(v_c, \vec{w}_c)$$

$$c \vec{w}_c' = g(v_c, \vec{w}_c) \ . \tag{4}$$

If we let

$$X = \begin{pmatrix} v_c' \\ v_c \\ \vec{w}_c \end{pmatrix}$$

then with appropriate identifications the equations (4) take the form

$$X' = F(X) = A_c X + h_c(X)$$

where the constant coefficient matrix $A_c = F'(0)$ is the linearization about the rest state.

Since the structure of the stable and unstable manifolds in phase space is a crucial feature in our analysis of this system, it is extremely useful to know that a consequence of the assumption that $(v, \vec{w}) = (0, 0)$ is a stable rest state of (1) is that [4] for $c > 0$ the matrix A_c has one positive eigenvalue and that the remaining eigenvalues have negative real parts. This is easily verified for the FitzHugh-Nagumo equations. The resulting one dimensional unstable manifold $\mathfrak{u}$ and n-dimensional stable manifold $\mathfrak{s}$ can be skematically represented as in Figure 1.

The class of traveling wave solutions can be broken down into several subclasses. We're most concerned here with an impulse solution (or homoclinic solution), that is, one which satisfies the condition

$$\lim_{\eta \to \pm\infty} (v_c(\eta), \vec{w}_c(\eta)) = 0 \quad .$$

Other classes of traveling wave solutions are those which approach the

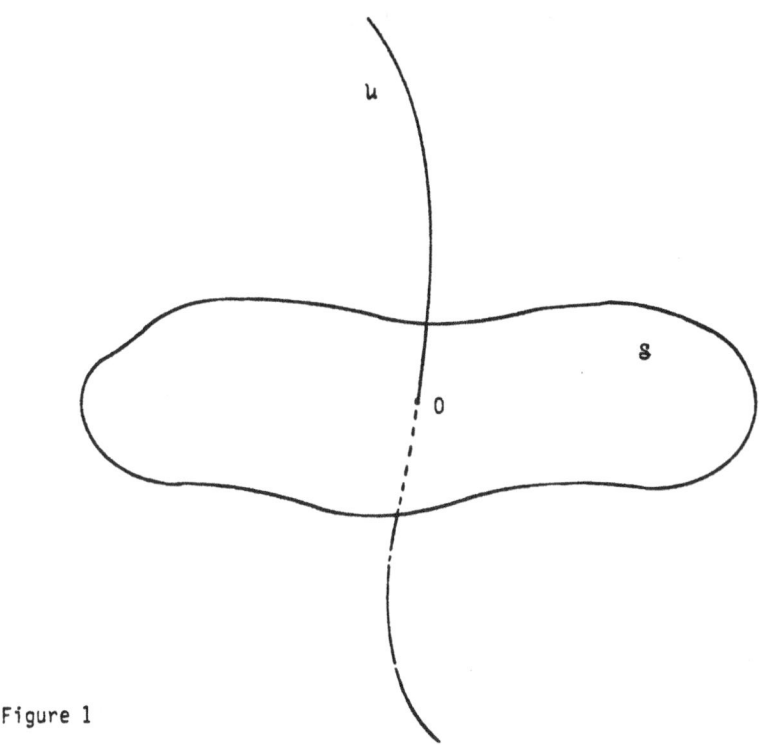

Figure 1

rest state in only one of the two directions and those which do not

converge to the rest state in either direction. This latter class

includes periodic solutions. For the FitzHugh-Nagumo equations the

existence of single pulse impulse solutions and periodic solutions is

well established [1, 3, 13, 14] . In addition it is known [9, 15] that

there exists both an infinite family of periodic and nonperiodic

solutions which do not converge to the rest state at all, as well as

a family of solutions which converge to the rest state only as
$\eta \to +\infty$.

The value of the parameter c determines whether or not an
impulse solution exists. Since an impulse solution must take off
from rest (that is, converge to 0 as $\eta \to -\infty$) there are only two
possible trajectories given any value of $c > 0$, namely the top and
bottom branches of the unstable manifold $\mathcal{U}$. For any of the standard
nerve models one of these (say the bottom) can never return to rest.
Since the solution also must return to rest (converge to 0 as
$\eta \to +\infty$), the trajectory of a solution with initial condition on $\mathcal{U}$
must then intersect $\mathcal{S}$. In general this occurs for isolated values of
c, say at some value c_0. Figure 2 illustrates typical behavior for
the phase space when c is near c_0.

It is interesting to note that whether the trajectory for $\mathcal{U}$
slices from above to below or from below to above as c increases
through c_0 is important to the question of the stability of the
resulting impulse solution as a solution to (1) [5, 7]. Existence
questions are independent of this feature and we'll work with the
arbitrary assumption that the passage is from below to above as shown.

The term "multiple pulse impulse solution" is somewhat ambiguous.
Intuitively one looks for a solution which arises from rest and then
loops through phase space two or more times before converging to rest,
with each loop returning near the rest state. See Figure 3. This may
be made precise by chosing a fixed neighborhood of the rest state and

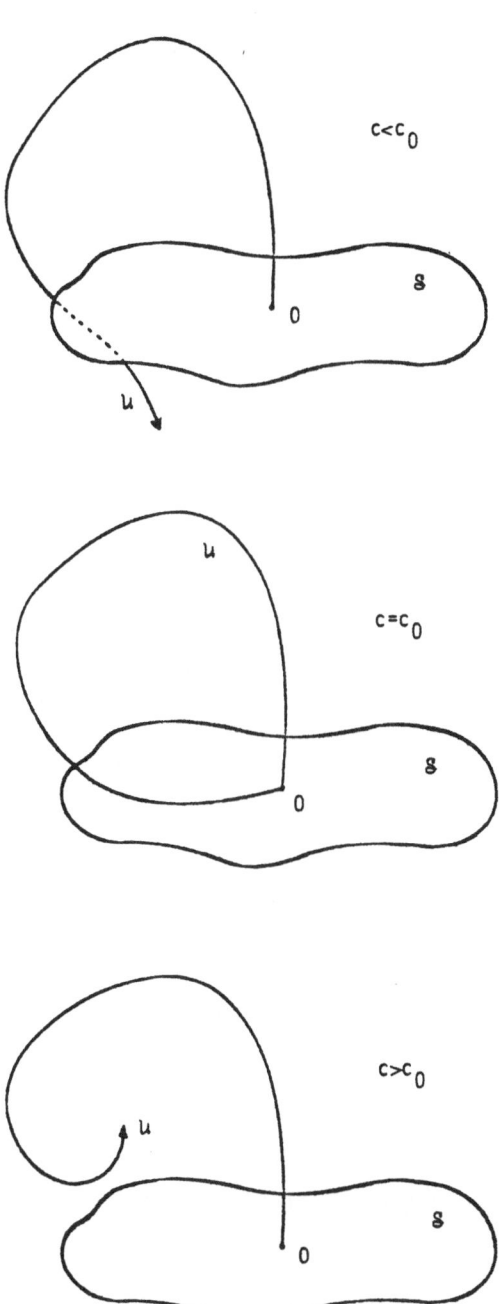

Figure 2

defining a solution to have k pulses if the trajectory exits (and
enters) that neighborhood k times. We will define such a
neighborhood. Note that by the choice of a small enough neighborhood
any impulse solution can be thought of as having a solitary pulse.
In nerve models the distinct spike of the action potential provides
a natural identification of pulse even if the mathematics is oblivious
to such considerations.

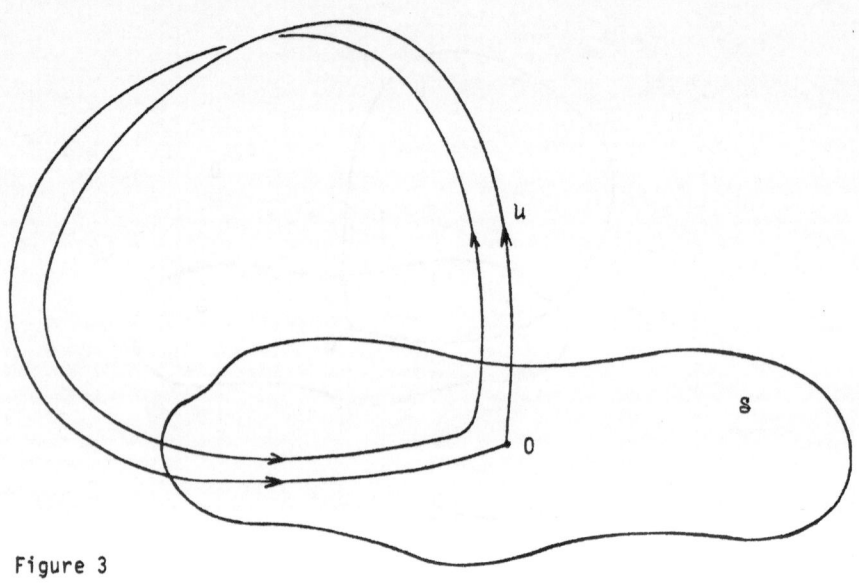

Figure 3

When dim $\vec{w} = 1$ (as in the FitzHugh-Nagumo equations) the conditions which produce multiple pulse solutions are

1) There exists $c_0 > 0$ such that (1) has an impulse solution when $c = c_0$

2) A_{c_0} has eigenvalues $\lambda > 0$ and $-\omega \pm \gamma i$ with $\gamma \neq 0$

(*) and $\lambda > \omega > 0$

3) A transversality condition which assures that as c passes through c_0 the crossing of the unstable manifold u through the stable manifold s (as in Figure 2) is not tangential.

For models with $\dim \vec{w} > 1$ an additional transversality condition is required and condition 2 is slightly more complicated. The reader is referred to [6] for details.

The result which will be extended here is [6]

Theorem (Evans, Fenichel, Feroe): Under conditions (*) there exists a monotone sequence $\{c_i\}_{i \in \mathbb{Z}^+}$ with $\lim_{i \to \infty} c_i = c_0$ such for each $c = c_i$ the system (1) has a multiple pulse impulse solution consisting of two pulses.

For such a solution (see Figures 11 - 15 in Section 5) the two separate pulses are almost identical in form. Moreover, the separation between the two pulses increases by approximately the amount π/γ with each unit increase of the index i in the velocity c_i. The computation of such solutions for the piecewise linear

FitzHugh-Nagumo equations is presented in Section 5 and in [10] along with a discussion of the stability properties of the solutions viewed as solutions of (1).

Hastings [15] has verified that for the cubic FitzHugh-Nagumo equations there are parameter values which satisfy conditions 1 and 2 of (*), and since condition 3 simply avoids exceptional cases, double pulse solutions almost certainly exist. It should also be noted that conditions for the existence of multiple pulse solutions are given by Carpenter [2], conditions which require that $\dim \vec{w} > 1$ and exploit quite different aspects of the geometry of the phase space.

To extend the analysis to trains of k pulses for $k \geq 2$, a more complicated indexing set is needed for the set of velocity values c which support multiple impulse solutions.

Definition: For $k \in \mathbb{Z}^+$ and $\rho > 0$ let

$$\Omega_k(\rho) = \left\{ (j_1, \ldots, j_{k-1}) \mid j_i \in \mathbb{Z}^+, \ \frac{1}{\rho} j_i < j_{i+1} < \rho j_1 \right\}$$

Note that when $k = 2$ (the number of pulses), $\Omega_2(\rho) = \mathbb{Z}^+$ the indexing set used previously for double pulses. To get the feel of $\Omega_k(\rho)$ the reader should consider the following examples:

$$(3,8,7) \in \Omega_4(3)$$

$$(3,8,9) \notin \Omega_4(3)$$

$$(3,8,2) \notin \Omega_4(3) \ .$$

The principal result can now be stated as

Theorem: Under conditions (*) there is a constant $\rho > 1$ such that for each $k \in \mathbb{Z}^+$ there exists a set of c values indexed by $\Omega_k(\rho)$ where for each such c value the system (1) has a multiple impulse solution consiting of k pulses.

As was the case previously the individual 'pulses all have similar forms. Also, the integers in the index $(j_1,\ldots,j_{k-1}) \in \Omega_k(\rho)$ indentify the spacing between pulses, with the space between the ℓ and $\ell + 1$ pulses increased by approximately the amount π/γ when j_ℓ is increased by one. The set of velocity values has a somewhat complicated nested structure.

3. Geormetric analysis

A. The System

In this section we assume that conditon (*) is satisfied by a system with dim $\vec{w} = 1$ such as the FitzHugh-Nagumo equations. The phase space is then 3 dimensional.

When $c = c_0$ the presence of complex eigenvalues for the matrix A_{c_0} causes a trajectory on the stable manifold $\mathcal{S}$ to spiral in towards the rest state as $\eta \to \infty$. Tne orbit of the solitary impulse is sketched in Figure 4. In addition in Figure 4 a "pill box" with flat top and circular sides has been constructed which sits on $\mathcal{S}$. Two "windows" have been identified on the surface of the pill box, one labled TOP through which the trajectory of the impulse solution rises as η increases, and one labled SIDE through which the trajectory crosses (on the lower boundary of SIDE) on its return to the rest state. For simplicity in a neighborhcod of rest we draw the unstable manifold $\mathcal{u}$ as a straight line (z axis) perpendicular to the stable manifold $\mathcal{S}$ which is drawn as a plane (x, y plane).

The situation when $c \geq c_0$ with c near c_0 is indicated in Figure 5. The flow of the system maps points on SIDE to TOP and points on TOP to SIDE. Local coordinates (ℓ, z) and (x,y) have been assigned to SIDE and TOP respectively with x, y and z retaining their original values as coordinates of 3-space. The rectangle SIDE is mapped to the spiral ribbon drawn in TOP. The dashed vertical line in SIDE is mapped

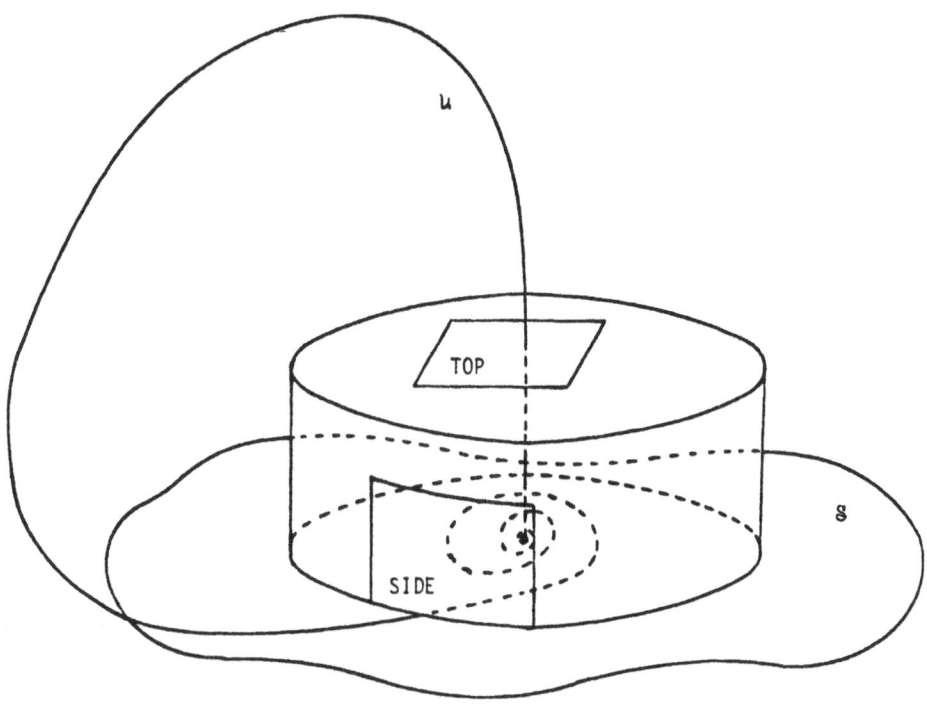

Figure 4

to the spiral dashed line in TOP and is included here simply to clarify the map. The spiral ribbon in TOP is in turn mapped to the spiral ribbon drawn in (overlapping, actually) SIDE.

Define $p_1(c)$ to be the first intersection of u with SIDE. In Figure 5 that point is drawn in SIDE as it might be for $c > c_0$. The spiral image drawn in SIDE also has $p_1(c)$ as its center. For $c = c_0$ the point $p_1(c)$ would be at $(z, \ell) = (0,0)$.

Of obvious concern here are those points of SIDE which are carried by this mapping of SIDE to TOP to SIDE to points of $\mathcal{g} \cap$ SIDE. Such points are the inverse images of the points drawn with bold lines in the spiral (Figure 5). Those inverse images are

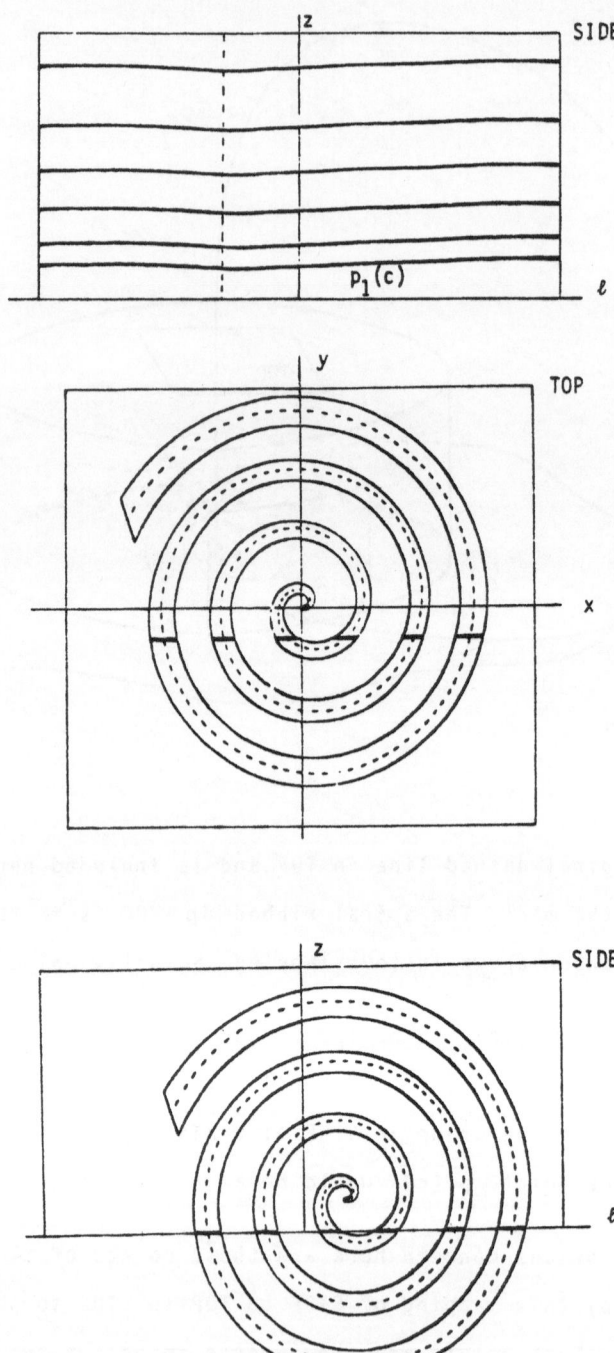

Figure 5

more or less horizontal lines as drawn with bold lines in the first drawing of SIDE in Figure 5, and will henceforth be referred to as "hot lines".

Consider what changes in Figure 5 occur as $c \to c_0$. The point $p_1(c)$ moves toward the point $(z, \ell) = (0, 0)$ and with it the center of the spiral image in SIDE. The shape of the spiral itself undergoes little change and consequently more and more of it intersects the $z = 0$ line (ie. the stable manifold s). This in turn produces more and more hot lines in SIDE which appear below the hot lines already there. Existing hot lines move very little as $c \to c_0$. Finally at $c = c_0$ as shown in Figure 6, there are an infinite number of hot lines converging exponentially to $z = 0$. Those lines can be numbered as indicated.

Note that a trajectory initially on line m spends more time in the pill box than one initially on line m' for $m > m'$. This observation becomes important when one considers that the time the trajectory spends in the pill box (ie. near rest) is the spacing between two pulses. That time is approximately proportional to $-\ln z$.

B. Interaction

In the following figures we look at the interaction of the hot line structure with the trajectory of the unstable manifold u as $c \to c_0$. The fact that any hot line moves very little once it appears (ie. for c close enough to c_0) permits us to treat each one as a

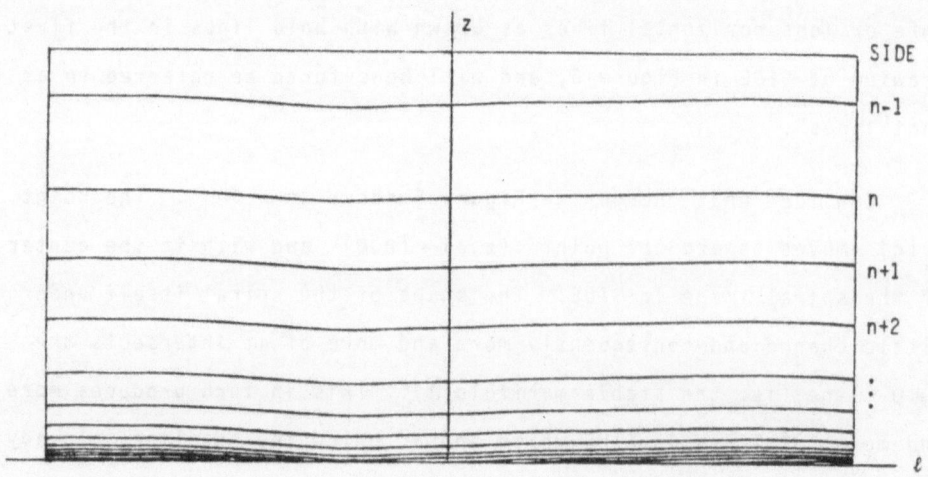

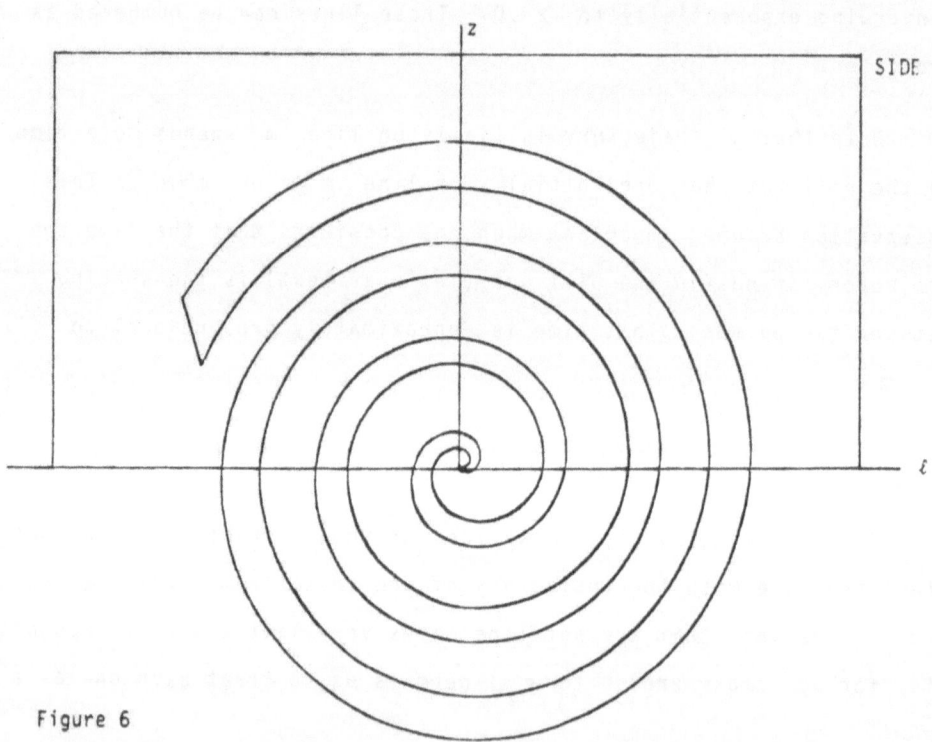

Figure 6

fixed line as we vary c .   Unlike previous figures where c was
fixed we now display the effects varying c .

The previously identified $p_1(c)$ traces a curve in SIDE as shown
in Figure 7. Figure 7 also displays a selection of (dashed) hot lines.
At the value c_n of c where the image of $p_1(c)$ intersects the $n\underline{\text{th}}$
hot line there is a double pulse solution. For that value c_n , the
trajectory μ arises from rest, passes up through TOP, returns to the
pill box at $p_1(c_n)$ in SIDE, passes up through TOP a second time,
again returns to SIDE this time on $\mathcal{S}$, and converges back to rest.

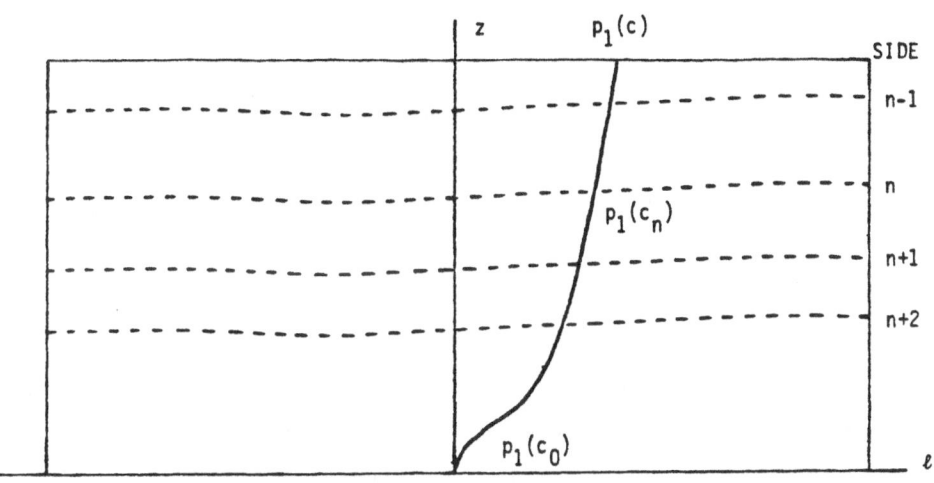

Figure 7

Obviously, to claim that double pulse impulse solutions exist, one needs to know that as $c \to c_0$ the emerging hot lines appear below the descending image of $p_1(c)$. Such behavior is produced by the part of condition (*) that requires $\lambda > \omega$. The Evans-Fenichel-Feroe result then follows with c_i being the velocity value such that $p_1(c_i)$ is on the $i^{\underline{th}}$ hot line.

The observation about increased spacing between pulses for increases in i follows from the fact that the spacing is determined by the time the trajectroy spends in the pill box. That time increases by approximately one half period of the rotating spiral (Figure 5), a value of about π/γ.

The observation that the separate pulses have approximately the same shape follows from the fact that the second exit of the trajectory from TOP is close to the first exit since while in the pill box the trajectory spirals in close to rest. Thus the two paths of the trajectory going from TOP to SIDE are similar and it is that portion of the trajectory where one sees a pulse shape. This approximation, as in the case for the statement about approximate spacing, becomes more precise as i increases. Both approximations can be expressed as limits.

The extension of the argument to triple pulses is studied by defining $p_2(c)$ to be the second intersection of u with SIDE. The image of $p_2(c)$ is shown in Figure 8 along with the image of $p_1(c)$. The dashed lines are again hot lines. The bold line section of the image of $p_1(c)$ corresponds to the bold line in the image of $p_2(c)$. Note that for $c = c_n$ the point $p_1(c_n)$ is on the nth hot line and the point $p_2(c_n)$ is on the ℓ axis.

A triple pulse exists for every value of c for which $p_2(c)$ intersects a hot line . Using the bold line section (Figure 8) of $p_2(c)$ as an illustration one can see that when c is near c_n so that $p_1(c)$ is near the nth hot line then the image of $p_2(c)$ will intersect some hot lines, but not all. It will miss hot lines with low numbers because the $p_2(c)$ spiral doesn't reach high enough in SIDE, and it will miss hot lines with high numbers because they have not yet emerged for those values of c . Both conditions depend on the value of n . Expanding the notation, let $j_i = n$, the number of the hot line near $p_1(c)$, and let j_2 be the number of a hot line intersected by $p_2(c)$ when c is near c_{j_1} . The above restriction on j_2 may be expressed as

$$\frac{1}{\rho} j_1 < j_2 < \rho j;$$

for some $\rho > 1$.

For such a permissible indexing pair (j_1, j_2), that is, for $(j_1, j_2) \in \Omega_2(\rho)$, there is a value c such that the trajectory of u intersects SIDE near the j_1 line on its first pass, on the j_2 line on its second pass, and on $\mathcal{S}$ on its third. In short, it is a triple pulse impulse solution. Again the spacing is seen to correspond to the values of j_1 (spacing between the first two pulses) and j_2 (spacing between the second two).

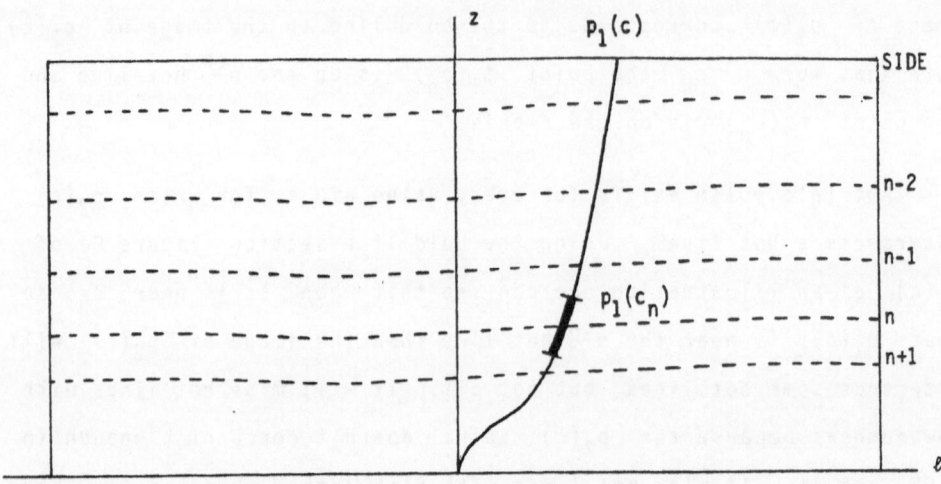

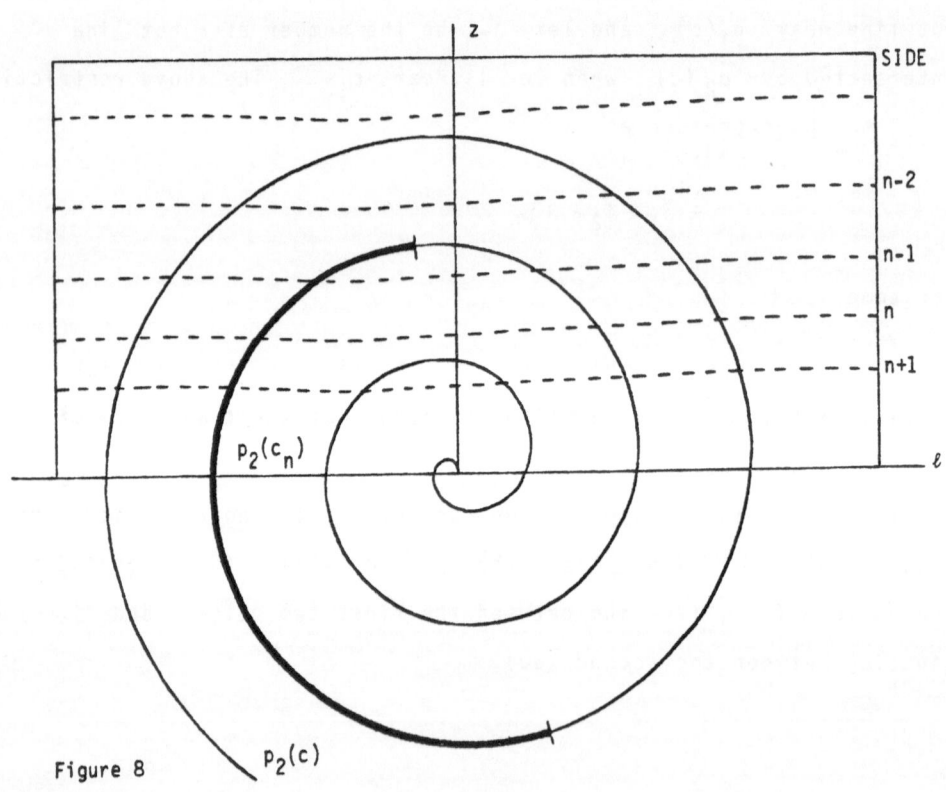

Figure 8

The case for k-pulse solutions for $k > 3$ is only slightly more complicated. Let $p_3(c)$ be the third intersection of u with SIDE. A portion of the image of $p_3(c)$ is shown in Figure 9 as a dotted line spiral with a bold line section. The portion that is drawn corresponds to that part of the image of $p_2(c)$ with the corresponding dotted and bold lines, with the bold line section near the j_2 hot line. This, in turn, is traced back to the dotted and bold line sections of $p_1(c)$, all near the j_1 hot line.

Impulse solutions with 4 pulses exist for values of c where $p_3(c)$ intersects a hot line. As before, hot lines with numbers which are too low or too high are missed. Those with too low numbers are missed if j_2 is too large, for then the spiral of $p_3(c)$ does not reach high enough in SIDE. Hot lines with numbers which are too high are missed because they just don't yet exist — a fact that depends on the value of j_1. This can all be combined into the condition that $p_3(c)$ intersects the j_3 hot line when

$$\frac{1}{\rho} j_2 < j_3 < \rho j_1$$

that is,

$$(j_1, j_2, j_3) \in \mathcal{R}_4(\rho)$$

The pattern for further extention to k-pulse solutions is the same as for 4-pulse. The geometric analysis of the principal result is complete.

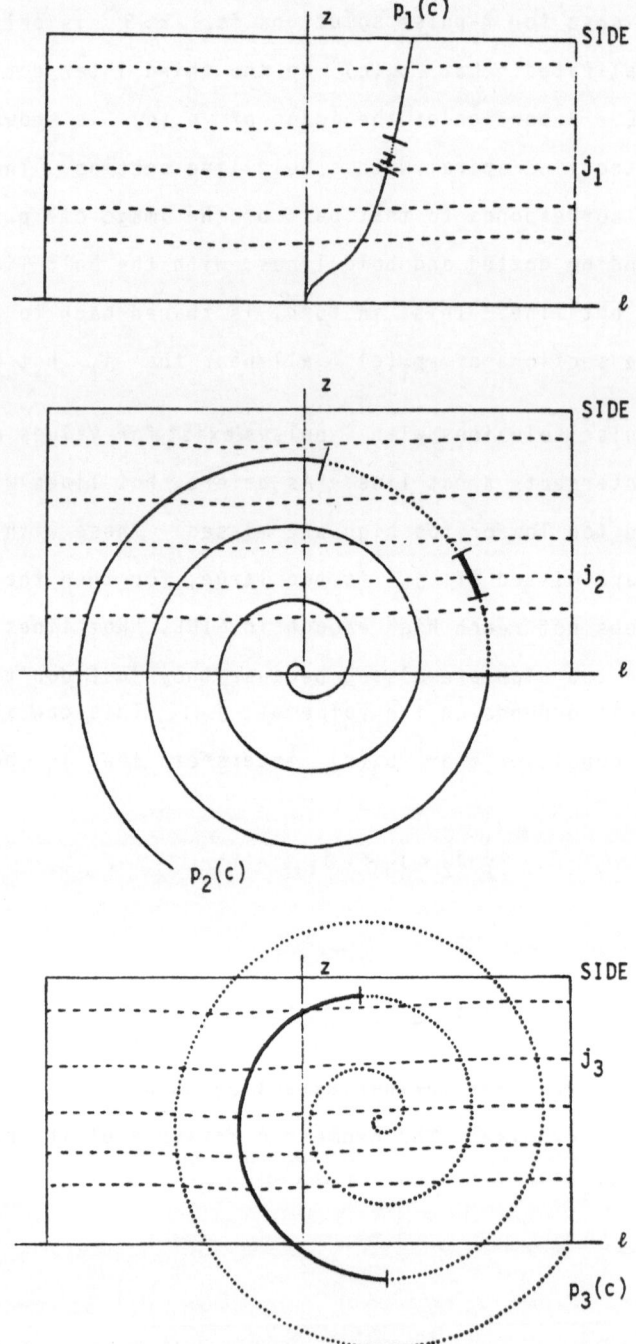

Figure 9

It should be noted that the problem being solved here is: Given an impulse solution with $k-1$ pulses, what are the restrictions on adding an additional pulse to get a solution with k pulses? This is different from the more general problem: Find all k-pulse solutions. For instance, by viewing a double pulse solution as a "two-humped solitary pulse", the Evans-Fenichel-Feroe result says that there are 4-pulse solutions consisting of 2 arbitrarily widely separated two-humped pulses. Such 4-pulse solutions do not all appear in this result.

4. The Analytic argument

A. The system

In this section we sketch the analytic realization of the
geometric motivation just presented. By restricting our attention
to the piecewise linear version of the FitzHugh-Nagumo equations
many of the technical difficulties are avoided.

The three dimensional phase space is divided in half by the
plane on one side of which the Heaviside step function is zero and
on the other side of which it is one. The origin (rest state) is
in the interior of the side where the Heaviside function is zero.
Thus the pill box can be constructed entirely within the region where
the system (4) is a first order constant coeficient linear three
dimensional homogeneous system.

A series of coordinate changes permits us to view the system (4)
as

$$
\begin{pmatrix} x(\eta) \\ y(\eta) \\ z(\eta) \end{pmatrix}' = \begin{pmatrix} -\alpha(s) & -\beta(s) & 0 \\ \beta(s) & -\alpha(s) & 0 \\ 0 & 0 & 1 \end{pmatrix} \begin{pmatrix} x(\eta) \\ y(\eta) \\ z(\eta) \end{pmatrix} \tag{5}
$$

within a pill box of height and radius one,

$$
\{(x,y,z) \mid x^2 + y^2 \leq 1 , \ 0 \leq z \leq 1\} .
$$

The necessary changes are:

i) The parameter c has been replaced by $s \overset{def}{=} -\ln(c - c_0)$.
The former concern for behavior as $c \to c_0$ is now for behavior
as $s \to \infty$.

ii) A linear change of coordinates results in the normal form
of the matrix A_c.

iii) A change in the independent variable results in the single
positive eigenvalue being one.

iv) The new complex eigenvalues $-\alpha(s) \pm i\beta(s)$ satisfy

$$\lim_{s \to \infty} -\alpha(s) \pm i\beta(s) = \bar{\alpha} \pm i\bar{\beta}$$

and the requirement that $0 < \omega < \lambda$ now becomes that $0 < \bar{\alpha} < 1$.

v) A dilation of coordinates results in having a homogeneous
system within the pill box of height and radius one.

Note that with these changes the z-axis is the unstable manifold u
and the x,y-plane is the stable manifold s .

Futher changes permit us to say that outside the pill box the
flow takes a neighborhood of $(0,0,1)$ in the top of the pill box (the
$z = 1$ plane) to a neighborhood of $(1,0,0)$ in the side (the $x^2 + y^2 = 1$
cylinder). This map is a diffeomorphism as long as trajectories are
transversal to the plane where the Heaviside step function switches
values. We assume that transversality.

In $\mathbb{R}^3$ more coordinate changes allow us to represent this diffeomorphism from TOP to SIDE

$$\begin{pmatrix} x \\ y \\ 1 \end{pmatrix} \rightarrow \begin{pmatrix} \cos \ell \\ \sin \ell \\ z \end{pmatrix}$$

where in local coordinates

$$\begin{pmatrix} \ell \\ z \end{pmatrix} = \left[\begin{pmatrix} b_{11} & b_{12} \\ b_{21} & 0 \end{pmatrix} \right] + G(x,y; s) \begin{pmatrix} x \\ y \end{pmatrix} + \begin{pmatrix} a(s) \\ e^{-s} \end{pmatrix} \qquad (6)$$

for $G(x,y;s)$ and $a(s)$ smooth functions satisfying

$$G(0,0 \; ; s) = 0$$

and

$$\lim_{s \to \infty} a(s) = 0 .$$

Note that in the local (ℓ,z) coordinates that $p_1(s)$, the first intersection of the trajectory of the unstable manifold with SIDE is

$$p_1(s) = \begin{pmatrix} a(s) \\ e^{-s} \end{pmatrix} .$$

The coordinate changes required to get (6) are as follows:

i) A rotation results in $b_{22} = 0$.

ii) An additional dilation results in the mapping of $(0,0,1)$ to $(1,0,0)$ when $s = \infty$.

iii) A reparametrization of s results in the z coordinate of $p_1(s)$ being e^{-s} .

The Equation (5) inside the pill box can be solved explicitly. For any choice of parameter s the map from $SIDE \to TOP \to SIDE$ is

$$\begin{pmatrix} \cos \ell \\ \sin \ell \\ e^{-r} \end{pmatrix} \to \begin{pmatrix} e^{-a(s)} & \cos(\ell + \beta(s)r) \\ e^{-a(s)} & \sin(\ell + \beta(s)r) \\ & 1 \end{pmatrix} \to \begin{pmatrix} \cos L \\ \sin L \\ Z \end{pmatrix}$$

where in local SIDE coordinates the $SIDE \to SIDE$ map $(\ell, e^{-r}) \to (L, Z)$ is

$$\begin{pmatrix} L \\ Z \end{pmatrix} = \begin{pmatrix} L(r, \ell; s) \\ b_{12} \, e^{-a(s)r} \cos(\ell + \beta(s)r) + e^{-s} + e^{-a(s)r} R(r, \ell; s) \end{pmatrix}. \quad (7)$$

The precise form of $L(r, \ell; s)$ and $R(r, \ell; s)$ are not important, rather we only need to know that both are small for large r and s and small ℓ.

To characterize the hot lines for any fixed choice of s it is necessary to determine those values of the local coordinates (r, ℓ) which result in $Z = 0$ in Equation (7). That condition can be written as

$$\cos(\ell + \beta(s)r) + \frac{1}{b_{21}} e^{-s + a(s)r} + R(r, \ell; s) = 0 \quad (8)$$

or by letting $F(r, \ell; s)$ be the left hand side of (8) as

$$F(r, \ell; s) = 0$$

The crucial observation about (8) is that as r increases,

$\cos(\ell + \beta(s)r)$ oscillates between 1 and -1 and that as long as $r < s/\bar{a}$ and s and r are both large, then the second and third terms are both small. By the intermediate value theorem, if s is large enough then several solutions $r(\ell,s)$ exist where

$$F(r(\ell,s),\ell;s) = 0 .$$

Each such solution can be numbered $r_n(\ell,s)$ where

$$n\pi/\bar{\beta} \le r_n(\ell,s) \le (n+1)\pi/\bar{\beta} ,$$

that is, in SIDE the $n^{\underline{th}}$ hot line is found in the horizontal strip

$$e^{-(n+1)\pi/\bar{\beta}} \le z \le e^{-n\pi/\bar{\beta}} .$$

The implicit function theorem implies that for each n the function $r_n(\ell,s)$ is unique and differentiable.

B.   Interaction

With the characterization of hot lines complete, we turn our attention to the existence of multiple pulse solutions through the successive intersection of ʯ with hot lines in SIDE as we vary s. The z coordinate of the first intersection $p_1(s)$ is $z_1(s) = e^{-s}$ a value which is greater than $e^{-s/\bar{a}}$. Since hot lines exist in the region of SIDE above the line $z = e^{-s/\bar{a}}$ we know that those line have already formed below the descending (as s increases) image $p_1(s)$ and that consequently $p_1(s)$ will intersect each hot line. The points of intersection identify double pulse solutions.

Triple pulse solutions exist because $z_2(s)$, the z coordinate of $p_2(s)$, satisfies

$$z_2(s) = b_{21} e^{-a(s)} \cos(\beta(s)s + a(s)) + e^{-s} + e^{-a(s)} R(s, a(s); s) > e^{-r} \quad (9)$$

for some values of s and r. In geometric terms, the spiraling image of $p_2(s)$ rises above the line $z = e^{-r}$ and if hot lines exist at such r values then the trajectory must intersect them. Inequality (9) can be written as

$$\cos(\beta(s)s + a(s)) > \frac{1}{b_{21}} e^{-r + a(s)s} - \frac{1}{b_{21}} e^{-s + a(s)s} + \frac{1}{b_{21}} R(s, a(s); s)$$

which is satisfied for some s values as long as the right hand side is small. This is the case if $s < r/\bar{a}$ and s is large. This condition that $r < s/\bar{a}$ and the previous condition for the existence of hot lines, namely $s < r/\bar{a}$ combine to become

$$\bar{a} s < r < s/\bar{a} \quad (10)$$

The importance of (10) is that if s values are chosen which keep $p_1(s)$ in the j_1 strip in SIDE, that is, for s near $j_1 \pi/\bar{\beta}$, then $p_2(s)$ moves through several strips and therefore intersects several hot lines. In particular it intersects the j_2 hot line if $r = j_2 \pi/\bar{\beta}$ satisfies (10). In terms of the numbering system (10) becomes

$$\bar{a} j_1 < j_2 < j_1/\bar{a} \quad (11)$$

Or in short, $(j_1, j_2) \in \mathcal{Q}_3(\rho)$ for $\rho = 1/\bar{a}$. This is the condition identified in the theorem. Actually, to be precise, the jump from condition (10) to condition (11) involves a slight adjustment of the value of $\bar{a}$ so that the choice of ρ is some $1 < \rho < 1/\bar{a}$.

The extention to k pulses is much the same and will not be presented here.

5. Calculations

Figures 10 through 21 are computer generated graphs of $v_c(\eta)$ vs. η for the piecewise linear FitzHugh-Nagumo equations. The parameter values for all figures are $a = b = .2$ and $d = .05$. The specific information for each figure is given below.

Figure	No. of Pulses	$\Omega_k(\rho)$ Index	c(approx.)
10	1	0	.41576105
11	2	1	.41257666
12	2	2	.41576084
13	2	3	$c_0 - c_3 \approx 10^{-12}$
14	2	4	$c_0 - c_4 \approx 10^{-17}$
15	2	5	$c_0 - c_5 \approx 10^{-22}$
16	3	(1,1)	.41246889
17	3	(1,2)	.41257656
18	4	(1,1,1)	.41246705
19	4	(1,1,2)	$c_{11} - c_{112} \approx 10^{-9}$
20	4	(1,2,1)	.41257664
21	4	(1,2,2)	$c_{122} - c_{12} \approx 10^{-11}$

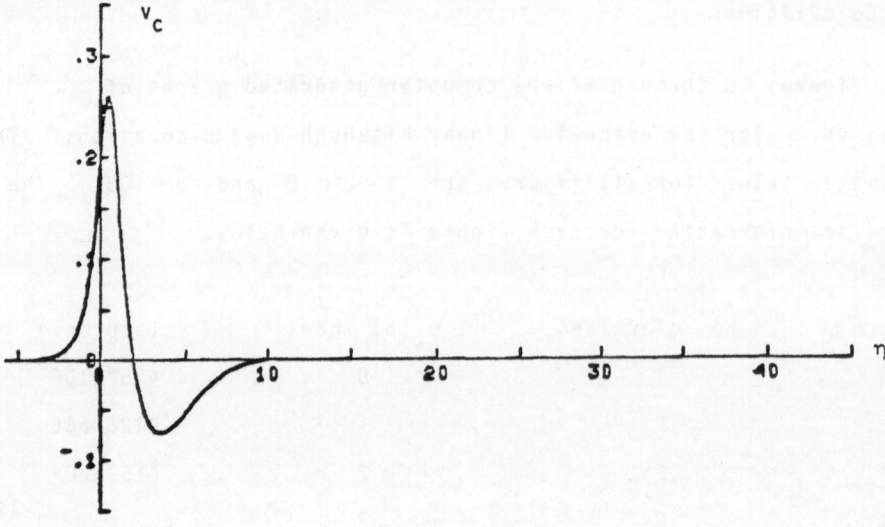

Figure 10

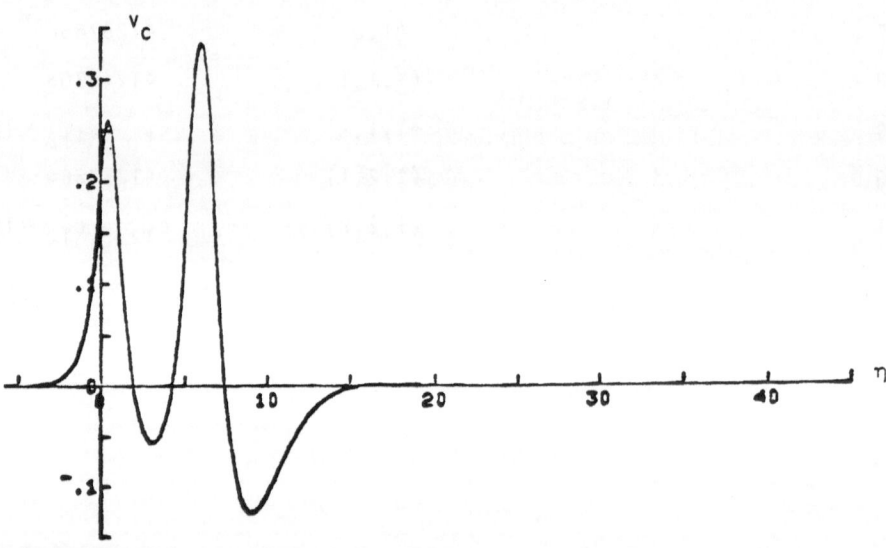

Figure 11

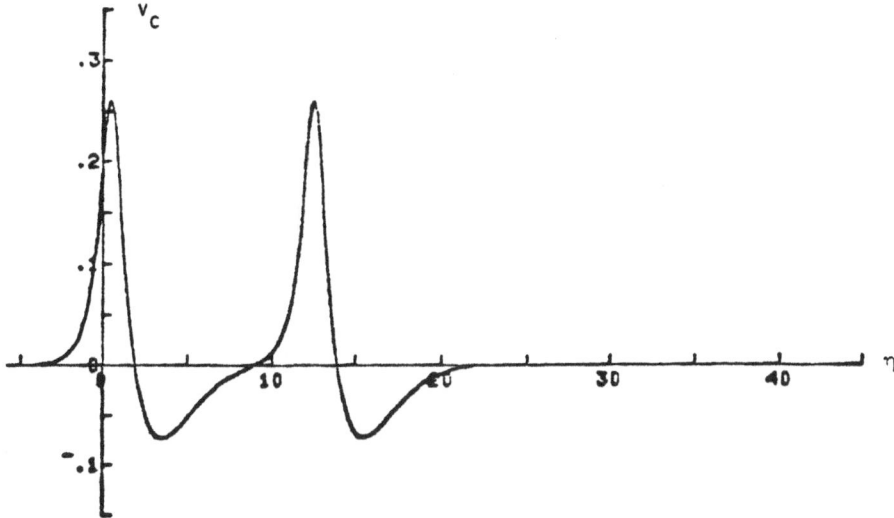

Figure 12

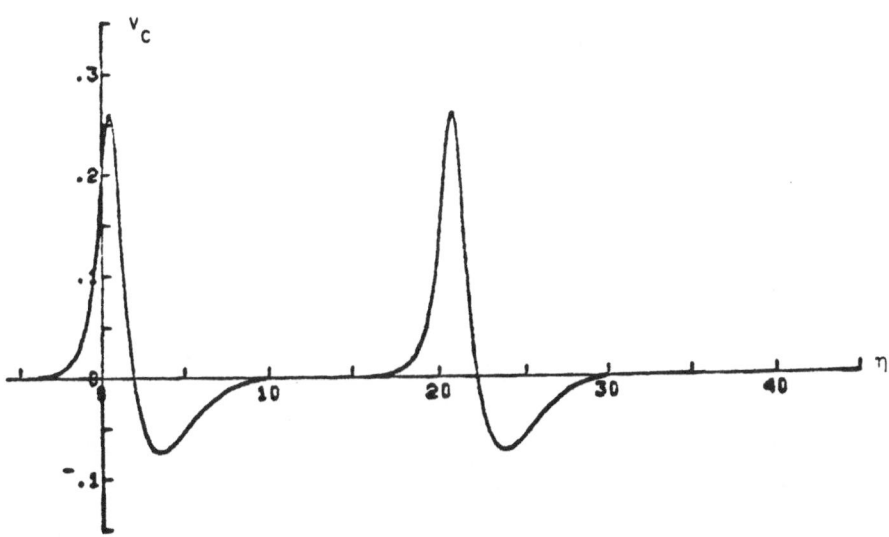

Figure 13

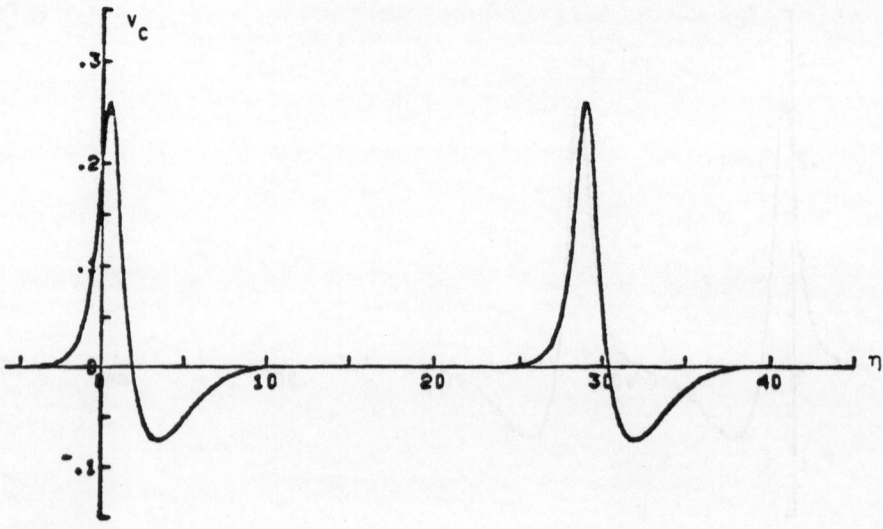

Figure 14

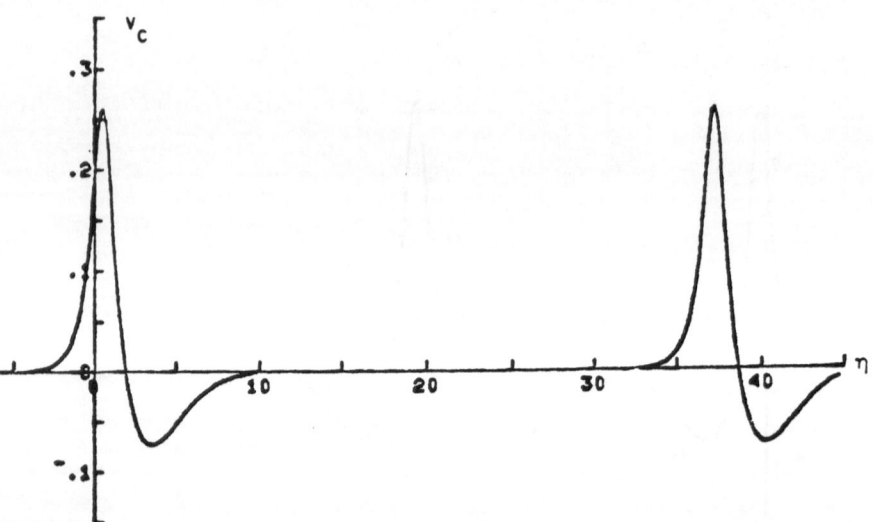

Figure 15

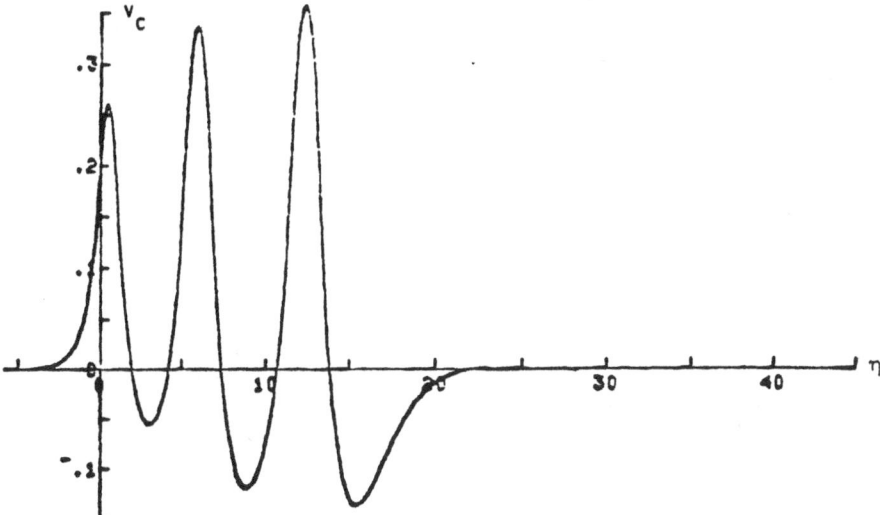

Figure 16

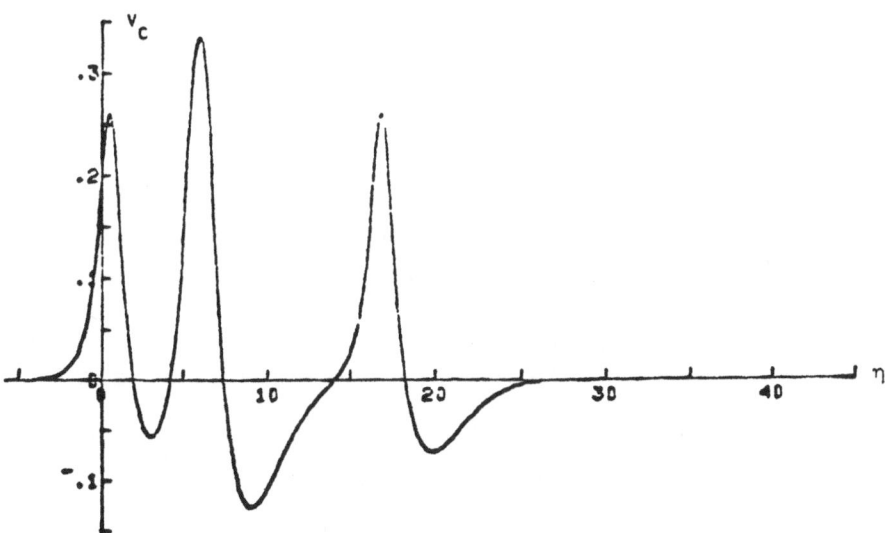

Figure 17

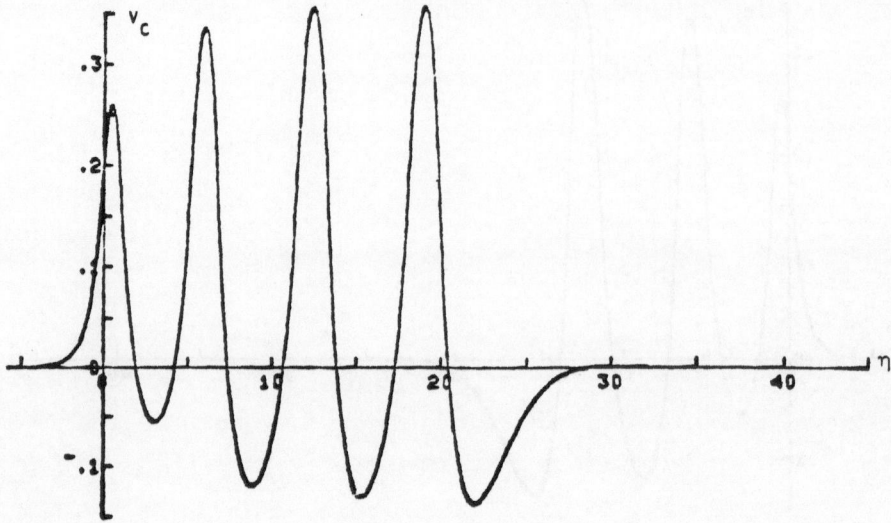

Figure 18

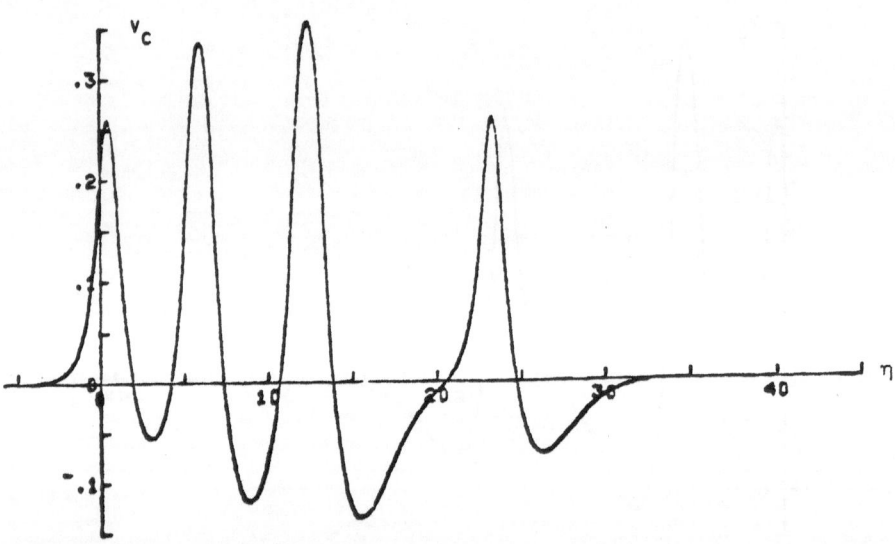

Figure 19

99

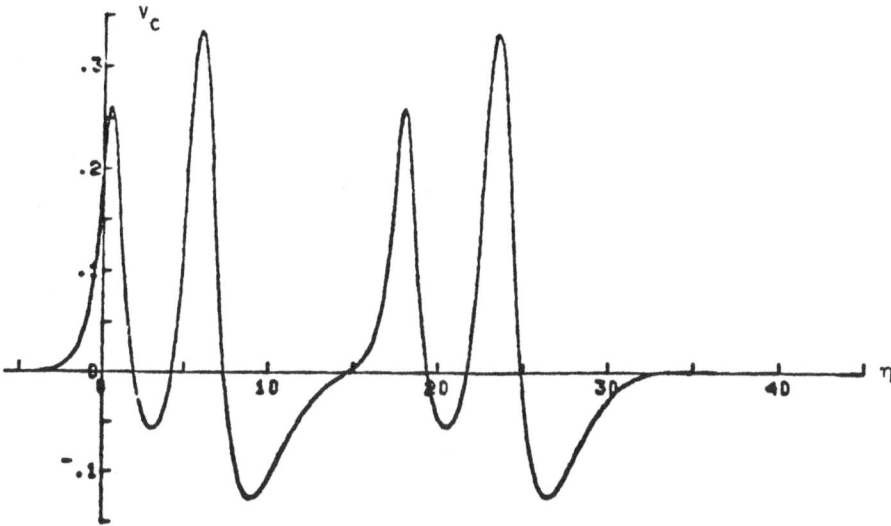

Figure 20

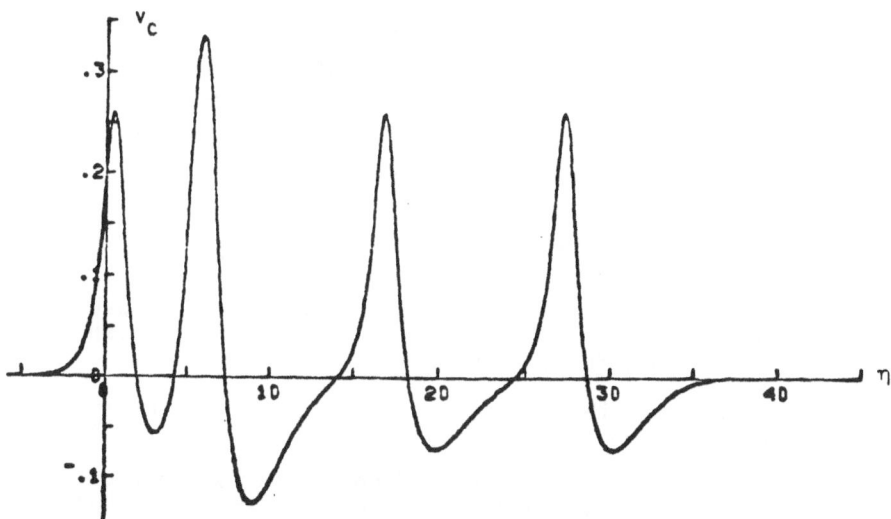

Figure 21

References

1. G. A. Carpenter, A geometric approach to singular perturbation problems with applications to nerve impulse equations, *J. Differential Equations*, <u>23</u> (1977), 335-367.

2. G. A. Carpenter, Burting phenomena in excitable membranes, *SIAM J. Appl. Math.*, <u>36</u> (1979), 334-372.

3. C. Conley, *Traveling Wave Solutions of Nonlinear Diffusion Equations*, Springer Lecture Notes in Physics, No. 38, Springer-Verlag, New York, 1975.

4. J. W. Evans, Nerve axon equations II. Stability at rest, *Indiana Univ. Math. J.*, <u>22</u> (1972), 75-90.

5. J. W. Evans, Nerve axon equations III. Stability of the nerve impulse, *Indiana Univ. Math. J.*, <u>22</u> (1972), 577-593.

6. J. W. Evans, N. Fenichel and J. Feroe, Double impulse solutions in nerve axon equations, *SIAM J. Appl. Math.*, <u>42</u> (1982), 219-234.

7. J. W. Evans and J. A. Feroe, Local Stability of the nerve impulse, *Math. Biosci.*, <u>37</u> (1977), 23-50.

8. J. A. Feroe, Temporal stability of solitary impulse solutions of a nerve equation, *Biophys. J.*, <u>21</u> (1978), 103-110.

9. J. A. Feroe, Traveling waves of infinitely many pulses in nerve equations, *Math. Biosci.*, <u>55</u> (1981), 189-203.

10. J. A. Feroe, Existence and stability of multiple impulse solutions of a nerve equation, *SIAM J. Appl. Math.*, <u>42</u> (1982), 235-246.

11. R. FitzHugh, Impulses and psysiological states in models of nerve membrane, *Biophys. J.*, <u>1</u> (1961), 445-466.

12. R. FitzHugh, Mathematical models of excitation and propagation in nerve, in *Biological Engineering*, H. P. Schwan, ed., McGraw-Hill, New York, 1969, Chapter 1, 1-85.

13. S. P. Hastings, The existence of periodic solutions to Nagumo's equation, *Quart. J. Math.*, (3) <u>25</u> (1974), 369-378.

14. S. P. Hastings, The existence of homoclinic and periodic orbits for the FitzHugh-Nagumo equations, *Quart. J. Math.*, (2) <u>27</u> (1976), 123-134.

15. S. P. Hastings, Single and multiple pulse waves for the FitzHugh-Nagumo equations, *SIAM J. MATH.*, <u>42</u> (1982), 247-260.

16. A. L. Hodgkin and A. F. Huxley, A qualitative description of membrance current and its application to conduction and nerve, *J. Physiol.* <u>117</u> (1952), 500-544.

17. H. P. McKean, Nagumo's equation, *Adv. Math.*, <u>4</u> (1970), 209-223.

18. J. Nagumo, S. Arimoto, and S. Yoshizawa, An active pulse transmission line simulating nerve axon, *Proc. I.R.E.*, <u>50</u> (1962), 2061-2070.

19. J. Rinzel and J. B. Keller, Traveling wave solutions of a nerve conduction equation, *Biophys. J.*, <u>13</u> (1973), 1313-1337.

20. L. P. Sil'nikov, A contribution to the problem of the structure of an extended neighborhood of a rough equilibrium state of saddle-focus type, *Math. USSR-Sb.*, <u>10</u> (1970), 91-102.

DYNAMIC MODELS OF NEURAL SYSTEMS:

PROPAGATED SIGNALS, PHOTORECEPTOR

TRANSDUCTION, AND CIRCADIAN RHYTHMS

by

Gail A. Carpenter[*]
Department of Mathematics
Northeastern University
Boston, Mass.   02115

and

Center for Adaptive Systems
Department of Mathematics
Boston University
Boston, Mass.   02215

and

Stephen Grossberg[**]
Center for Adaptive Systems
Department of Mathematics
Boston University
Boston, Mass.   02215

[*]   Supported in part by the Air Force Office of Scientific Research
      (AFOSR 82-0148), the National Science Foundation (MCS-80-04021),
      and the Northeastern University Research and Scholarship Develop-
      ment Fund.

[**] Supported in part by the Air Force Office of Scientific Research
      (AFOSR 82-0148) and the National Science Foundation (NSF ISF-80-
      00257).

TABLE OF CONTENTS

INTRODUCTION

This article presents dynamical system models of three types of related neural phenomena: electrical signal patterns in individual nerves, transduction of light into electrical signals by photoreceptors, and the electrical, chemical, and light interactions that control circadian rhythms.

Each of these phenomena takes place in a complex neural system, which can be experimentally analyzed by a variety of techniques, each technique probing different levels of system organization; e.g., behavioral, neurophysiological, anatomical, and molecular. Our modelling approach aims to discover and to classify system properties that will persist as new experimental methods are developed. To achieve this goal, each model mechanizes basic principles of neural design which we suggest are rate-limiting in the data gathered by any of several methods. To test whether these principles are operative in the cases we discuss, each model is used to predict how several data indices will simultaneously vary in response to prescribed parameter changes in each model. These predictions should hold not only in the cases treated herein, but also in all neural systems where these principles are rate-limiting. The predictions are in this sense general invariants that can be sought in any body of neural data.

In this spirit, we examine detailed parametric properties of solutions of differential equation models, both analytically and numerically, and compare these properties with related data. By contrast, various other approaches have sought the existence of solutions to justify a model, but have not discussed the detailed parametric structure of these solutions. For example, there has recently been great interest in the complex dynamics which can arise in such simple systems as maps of the interval (Feigenbaum, 1978; Li and Yorke, 1975). The combination of simple equations and complex dynamics has made these systems appealing candidates for biological models (Guevara et al, 1981; May, 1976). Our models are also capable of generating complex, even chaotic waveforms, but we believe that the predictable parametric regularities of the model solutions provide the strongest argument for their physical relevance.

I. SIGNAL PATTERNS IN SINGLE NERVE CELLS

I.1. Parametric Classification of Signal Patterns

In this section we examine a class of dynamical systems that generalizes the famous Hodgkin-Huxley (1952) model of nerve impulse propagation. Despite the many variations that exist across nerve cells and, more generally, excitable membranes throughout phylogeny, the intuitive concepts of ionic interactions with membrane voltage that led to the Hodgkin-Huxley (1952) model have proved to be universal. The generalized Hodgkin-Huxley model attempts to rigorously capture both the invariant intuition and the many variations on the ionic hypothesis. To structure some of these experimental variations and to test whether Hodgkin-Huxley dynamics generate the observed signal patterns, the detailed parametric properties of model solutions have been examined and classified (Carpenter, 1977a, 1977b, 1979, 1981).

One surprising result of this analysis is that the mere existence of the elementary components of impulse propagation (Na^+ entering the cell followed by K^+ leaving the cell) implies many properties which had previously been ascribed to additional membrane processes. For example, bursting patterns measured from epileptic neurons have been ascribed to a complex interaction between neurons (Ward, 1969), but these bursting patterns can be generated by individual neurons (Carpenter, 1979, 1981). Other characteristics of the data had not even been noticed. For example, there exist two types of regular periodic (beating) signal patterns, each type possessing a series of correlated distinguishing properties. The model hereby structures signal patterns that had previously seemed to be so irregular that their fine structure was ignored. In this sense, the model helps to define what the reliable data properties are by parametrically relating these properties to a single underlying mechanism.

In this article, a number of parametric model predictions will be presented. It is important to emphasize what we mean by a prediction. A prediction is a property of the model in question. If one of these properties fails to hold in vivo, either the model is inapplicable or there are additional factors at work. We will consider examples of data which do not correspond to the predicted solution types of one Hodgkin-Huxley model, but do correspond to the predicted solution types of a related Hodgkin-Huxley model that possesses one more ionic process. An important goal of the classification theory is to discover the minimal number and type of ionic processes that are needed to generate prescribed signal patterns. From this perspective, the popular FitzHugh-Nagumo model (Evans, Fenischel, and Feroe, 1982; FitzHugh, 1961; Hastings, 1982; McKean, 1970; Nagumo, Arimoto, and Yoshizawa, 1962; Rinzel and Keller, 1973) is a variant of a generalized Hodgkin-Huxley model with one ionic process less than the original Hodgkin-Huxley (1952) model.

I.2. The Hodgkin-Huxley Equations

The original Hodgkin-Huxley (1952) model was derived from experimental studies of the squid giant axon. The axon of a nerve cell is a long cylindrical process that leads from the nerve cell body to other cells or muscles (Figure 1). Propagated signals can hereby be transmitted along axons between communicating cells. The Hodgkin-Huxley

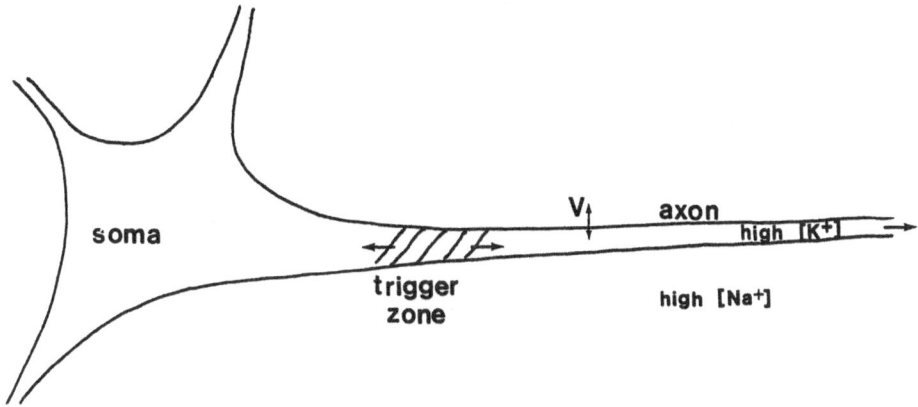

Figure 1: Schematic view of part of a nerve cell.

model describes interactions between membrane voltage V and three
ionic variables m, n, and h capable of generating these propagated
signals.

Denoting the distance traveled from the nerve cell body down the
axon by the variable x, the Hodgkin-Huxley (1952) equations describe
how the ionic processes m(x,t), n(x,t), and h(x,t) interact with the
voltage V(x,t) at each position x and time t, and how the voltages at
nearby positions influence each other via diffusion. The equation
governing the voltage V(x,t) is the membrane equation

$$\frac{a}{2R} \frac{\partial^2 V}{\partial x^2} = C\frac{\partial V}{\partial t} + g(V,m,n,h) \tag{1}$$

where the term $\frac{a}{2R} \frac{\partial^2 V}{\partial x^2}$ is the total membrane current density (by Ohm's
law) and the term $C\frac{\partial V}{\partial t}$ is the capacitance current density. These two

terms control the diffusion of voltage between spatial positions. The remaining term $g(V,m,n,h)$ is the total ionic current density, which is defined by Hodgkin and Huxley (1952) as

$$g(V,m,n,h) = \bar{g}_{Na}m^3h(V-V_{Na}) + \bar{g}_K n^4(V-V_K) + \bar{g}_L(V-V_L).$$ (2)

Each summand in (2) is a product of a conductance times a voltage difference. Term $\bar{g}_{Na}m^3h(V-V_{Na})$ is the (inward) sodium current density; term $\bar{g}_K n^4(V-V_K)$ is the (outward) potassium current density; and $\bar{g}_L(V-V_L)$ is a leakage current density. The main step in generalizing the Hodgkin-Huxley equations is to consider total ionic currents more general in form than (2) that may include fewer or more ionic currents, and to characterize the qualitative properties of those functions that control signal properties.

The voltage is coupled to the ionic processes via equations of the form

$$\frac{\partial m}{\partial t} = \gamma_m(V)(m_\infty(V)-m)$$ (3)

$$\frac{\partial n}{\partial t} = \gamma_n(V)(n_\infty(V)-n)$$ (4)

and

$$\frac{\partial h}{\partial t} = \gamma_h(V)(h_\infty(V)-h).$$ (5)

Each of the ionic equations is of the same general form. Equation (3), for example, possesses a positive voltage-dependent rate term $\gamma_m(V)$ and a positive voltage-dependent asymptote $m_\infty(V)$ to which m is attracted. The ionic equations differ in two basic ways. Some ionic processes respond quickly to voltage changes, others slowly. Some ionic processes increase with voltage increments, others decrease. Since m responds quickly relative to n and h, $\gamma_m(V)$ is large relative to $\gamma_n(V)$ and $\gamma_h(V)$. Since m and n tend to increase whereas h tends to decrease as V increases, $m_\infty(V)$ and $n_\infty(V)$ are increasing functions of V whereas $h_\infty(V)$ is a decreasing function of V. The choice of fast-slow and on-off distinctions are characteristic of the qualitative hypotheses that define the class of generalized Hodgkin-Huxley models and distinguish one model from another.

I.3. Propagated Signals and Traveling Waves

To study signals propagated along an axon, solutions are sought of the form

$$V(x,t) = V(s) \qquad (6)$$

where

$$s = x + \omega t. \qquad (7)$$

In other words, one seeks solutions which propagate down the axon (x) through time (t) at a speed (ω). Such solutions are called traveling waves. When (6) holds, equation (1) can be rewritten as the pair of equations

$$\frac{dV}{ds} = W \qquad (8)$$

and

$$\frac{dW}{ds} = \theta W + g(V,m,n,h) \qquad (9)$$

in terms of the new independent variable s and the parameter

$$\theta = \frac{2RC\omega}{a}. \qquad (10)$$

To emphasize the fact that m responds quickly whereas n and h respond slowly to fluctuations in V, we can redefine the voltage-dependent rate functions and rewrite equations (3)-(5) in terms of the new independent variable s as

$$\frac{dm}{ds} = \delta^{-1}\gamma_m(V)\,(m_\infty(V)-m) \tag{11}$$

$$\frac{dn}{ds} = \varepsilon\gamma_n(V)\,(n_\infty(V)-n) \tag{12}$$

and

$$\frac{dh}{ds} = \varepsilon\gamma_h(V)\,(h_\infty(V)-h)\,. \tag{13}$$

Since both δ and ε are assumed to be small, the rate function $\delta^{-1}\gamma_m(V)$ is large compared to the rate functions $\varepsilon\gamma_n(V)$ and $\varepsilon\gamma_h(V)$.

I.4. Bursts and Two Types of Regular Periodic Waves: Predictions and Data

Previous articles have developed the mathematical analysis of system (8)-(13) (Carpenter, 1977a,b, 1979, 1981). Here, we will focus on the detailed predictions of that analysis; compare the predicted solutions with experimental recordings; examine some types of signal patterns which are not consequences of the basic sodium-potassium mechanisms; and consider ways in which the model may be augmented to account for these patterns.

First we will consider experimental and mathematical evidence for the existence of two types of regular periodic waveforms and how these are related to periodic bursts. Examples of these signal patterns can be seen in Figure 2, which shows recordings taken from two snail yellow cells (Benjamin, 1978). Each column in Figure 2 depicts signal patterns measured in a single cell. Each signal pattern in a given column occurs when the cell is maintained at a given level of hyperpolarization, which is denoted by i. Both cells emit regular periodic signals when i=0, as in the bottom row of Figure 2, but cell A has a higher frequency. As the cells are gradually hyperpolarized (i becomes negative), cell A moves through a region of high frequency

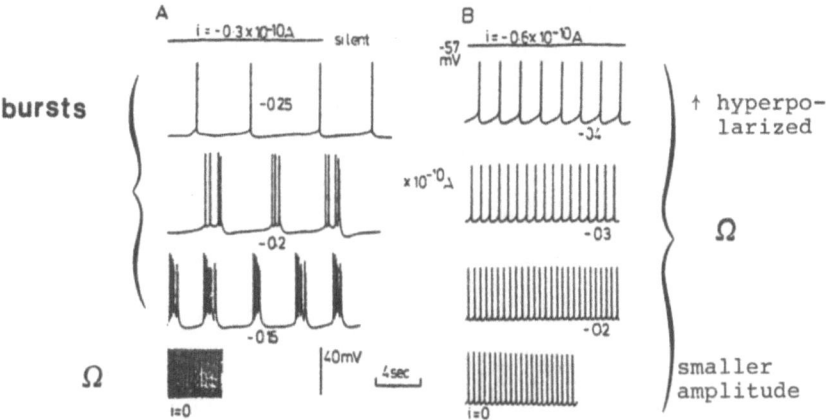

Figure 2: Recordings from two similar snail yellow cells (Benjamin,
1978, p.208), which illustrate the two types of predicted
dynamics of the generalized Hodgkin-Huxley model. From bottom to top, cells are hyperpolarized until they become
silent. (A) High frequency Ω-periodic spikes (i=0) pass
through a phase of bursts with many spikes, then bursts
with few spikes, then 1-bursts, then silent: exactly as
predicted. (B) In a cell without bursts (the other "half")
high frequency Ω-periodics become low frequency Ω-perio-
dics, then silent.

bursts, then low frequency bursts, then low frequency beats, then

becomes silent. Cell B continues to emit regular periodic signals

whose frequency declines gradually as i becomes negative until it,

too, is silent. Previous theorems (Carpenter, 1979, 1981) not only

predict the existence of the two types of signal patterns observed in

cells A and B, but also lead to further predictions, which cells A and

B confirm on closer inspection. Some of these predictions follow.

The most obvious difference between cells A and B in Figure 2 is

that cell A emits burst patterns at certain levels of hyperpolariza-

tion, but cell B never bursts. Until recently, it was not known that

Hodgkin-Huxley dynamics could lead to bursts at all. Surprisingly, it has been proved that "half" of all generalized Hodgkin-Huxley models, in a sense that can be made precise, admit bursting solutions (Carpenter, 1979). Given the unexpected ease with which bursts can be generated, it is imperative to study the detailed internal structure of these bursts to ascertain their underlying mechanism. It was also proved that all generalized Hodgkin-Huxley models admit regular periodic solutions, as do both cells A and B in Figure 2. To emphasize the meaning of these general results on bursts and periodic solutions, we mention that snail yellow cells, as in Figure 2, may sustain bursts in one season but only regular periodic patterns in another season (Benjamin, 1978). This can be explained by a suitable parameter shift in a generalized Hodgkin-Huxley model that removes the model from the parameter range where bursts occur to the parameter range where only regular periodic solutions occur.

The analysis of burst solutions leads to a geometric understanding of the phase portrait which is illustrated schematically in Figure 3. In Figure 3A, a burst with many spikes per burst, a so-called N-burst with N >> 1, is depicted. Each loop in the bursting trajectory corresponds to one spike in the cell's potential, as in Figure 2A. As more spikes in a burst unfold, the N-burst trajectory approaches a regular periodic solution that lies far from the equilibrium point (rest) in phase space. The regular periodic solution is called an Ω-periodic solution. When N is large, spikes late in the burst are all but indistinguishable from spikes in the Ω-periodic solution. Thus late in the burst, it appears as if the trajectory is approaching a limiting (ω-limit) set which in this case is the Ω-periodic solution. The part of the N-burst trajectory denoted by Q is a quiet spell during which the cell potential approaches close to equilibrium before the bursting cycle begins again.

The fact that an N-burst starts near rest and ends near an Ω-

A   B

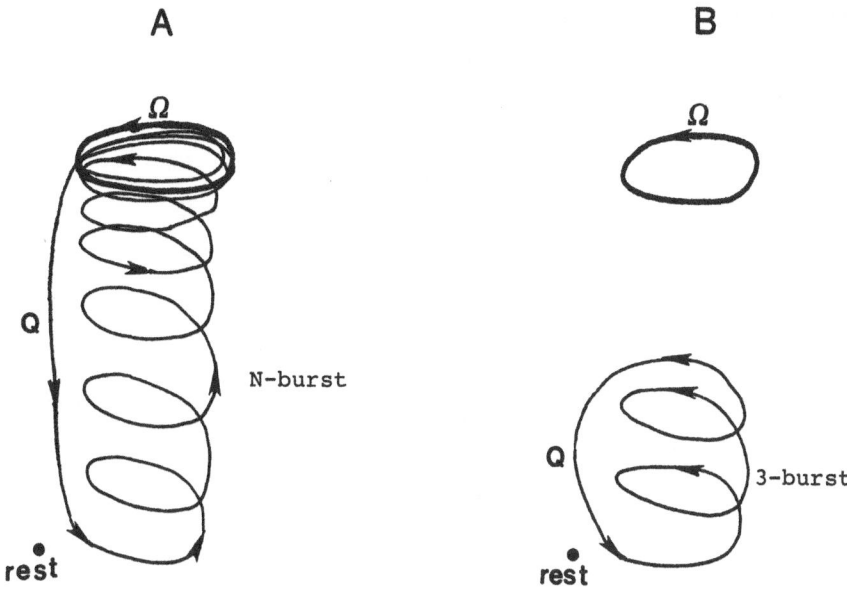

Figure 3:   Schematic representation, in phase space, of Ω-periodic
            solutions and (A) a burst with N spikes per burst; and
            (B) a burst with 3 spikes per burst. During the quiet
            spell (Ω) the burst solutions approach the rest point.

periodic solution far from rest has important implications for the
internal structure of each burst. Spikes emitted close to rest will
be emitted at a slow rate whereas spikes emitted close to the Ω-periodic solution far from rest will form a high frequency pattern of
approximately equally spaced spikes. Thus the spikes within a burst
will speed up until they abruptly shut off.

Figure 3B depicts a 3-burst. Since the last spike in this burst
is far from the Ω-periodic solution, all the spikes are emitted at a
lower rate than the final spikes in the N-burst of Figure 3A. Since
no spike in the 3-burst is close to the Ω-periodic solution, the regular frequency which the Ω-periodic solution imposes on the final

spikes in the N-burst need not appear in the 3-burst. Finally, compare the quiet spells Q in the N-burst of Figure 3A and the 3-burst of Figure 3B. Although both quiet spells correspond to a narrow range of subthreshold cell potentials, Figure 3 shows that they actually correspond to rather different paths in phase space. Consequently the durations of quiet spells between bursts with different numbers of spikes need not be the same. This point will be illustrated in Figure 9 below.

A generalized Hodgkin-Huxley model that possesses an N-burst solution also possesses a 1-burst, 2-burst, 3-burst,...., and an (N-1)-burst solution (Carpenter, 1979). Consequently, every generalized Hodgkin-Huxley model that admits bursts at all will admit 1-burst solutions. A 1-burst solution is a regular periodic solution (Figure 4). However, all generalized Hodgkin-Huxley models admit Ω-periodic

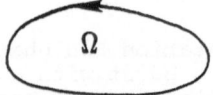

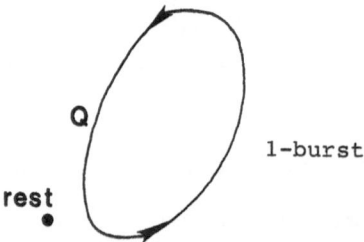

Figure 4: Schematic representation of the two types of regular periodic solutions whose existence is predicted in the "half" of the model cells which burst. The Ω-periodic is far from equilibrium while the 1-burst approaches equilibrium during the quiet spell.

solutions, which are also regular periodic solutions. Thus "half" of all generalized Hodgkin-Huxley models admit two classes of mechanistically distinct regular periodic solutions. How can an experimentalist know which regular periodic solution is being seen when a cell emits periodic spikes? To answer this question, the following parametric properties of each type of regular periodic solution may prove helpful.

First, in a model cell capable of bursting, the family of Ω-periodic solutions is always far from equilibrium, while each solution in the co-existing family of 1-burst solutions always approaches equilibrium during its quiet spell. Thus, we would expect that a small hyperpolarization of cell potential during a periodic 1-burst could easily extinguish this pattern, but a much larger hyperpolarization would be required to extinguish an Ω-periodic. Moreover, if an Ω-periodic pattern is hyperpolarized, it can become an N-burst solution. Given larger hyperpolarizations, N will tend to decrease. Given a large enough hyperpolarization, the Ω-periodic can be extinguished.

This prediction corresponds exactly to what occurs in Figure 2A, where successively larger hyperpolarizations transform a regular periodic pattern into bursting patterns with progressively fewer spikes per burst.

In generalized Hodgkin-Huxley models wherein bursts do not occur (the other "half" of the models), the regular periodic solutions that do occur are all Ω-periodic solutions. Hyperpolarization of such a solution moves it closer to rest, and thereby decreases the frequency of its spikes. If the hyperpolarization is chosen sufficiently large, the Ω-periodic solution is extinguished.

Figure 2B nicely illustrates this prediction. Successively greater hyperpolarizations cause progressively lower spiking frequencies, but do not cause bursts to occur, by contrast with Figure 2A.

Notice that the individual waveforms in Figure 2 differ in several ways which are hard to categorize <u>a priori</u>. For example, the waveforms corresponding to low frequency patterns in row 1 of the Figure are different. The periodic patterns of different frequency within column 1 and within column 2 of the Figure are different. Below, several of these differences will be parametrically characterized as typical predictions of Hodgkin-Huxley dynamics. It will also be shown how a single family of solutions in a model capable of bursting can possess properties from both column 1 and column 2 in a predictable order.

Because the underlying parameters of a given cell are not known a priori, the most robust predictions arise when the cell is parametrically perturbed in an experimentally controlled way, as in Figure 2A or 2B. Nonetheless it may be noted, as in row 1, that low frequency Ω-periodic solutions possess a strictly increasing ramp-like potential between successive spikes, whereas low frequency 1-bursts tend to be flat over a significant fraction of the interspike interval. Figure 2B illustrates the fact that Ω-periodic solutions of higher frequency tend to have smaller spike amplitudes. By contrast, the spike amplitudes of all bursts in Figure 2A are approximately constant in size. Finally, the spike amplitudes of all bursts exceed the spike amplitudes of high-frequency Ω-periodic solutions in a given model cell. In fact, the spike amplitudes of the 1-bursts in row 1 of Figure 2A are 50% larger than the amplitudes of the Ω-periodic spikes in row 4 of Figure 2A. These qualitative remarks are made more quantitative in the set of predictions depicted by Figures 5 and 6 below.

I.5. <u>A Single Family of 1-Bursts and Ω-Periodic Wave Trains</u>

A family of 1-bursts, as in Figure 2A, and a family of Ω-periodic solutions, as in Figure 2B, may or may not meet in phase space as changes in initial data generate successive members of the family

(Carpenter, 1979). Figure 5 depicts a family of solutions where the

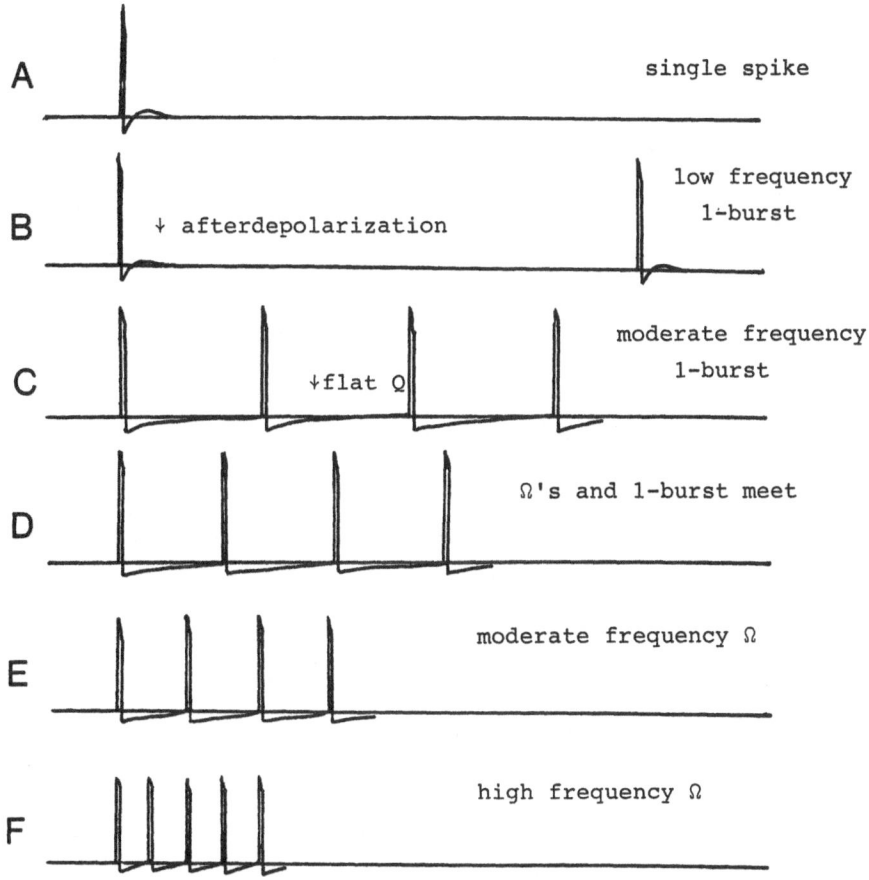

Figure 5: A typical family of regular periodic solutions of a generalized Hodgkin-Huxley model which has burst solutions. In
this example, the 1-burst family and the Ω-periodic family
meet, although this need not be the case. The quiet spell
goes from ∞ at the single pulse (A) to 0 at the dividing
point (D). The 1-burst frequencies range from very low to
moderate and the Ω-periodic frequencies range from moderate
to very high. The amplitudes of the 1-bursts ((A)-(D)) are
large and approximately constant over their wide frequency
range. In contrast, the amplitudes of the Ω-periodics de-
cline as the frequency increases, as does the size of the
hyperpolarization, or tail, following each spike.

two types of regular periodic patterns do meet. Figure 5A depicts a single spike, whose potential goes to the rest point both as t→+∞ and as t→-∞. Nearby in phase space is a 1-burst periodic solution (Figure 5B), which comes very close to the rest point during its quiet spells Q and thus has a low frequency. Such a low frequency 1-burst may or may not be followed by a brief afterdepolarization (arrow; also see Figure 8B). As the family is perturbed away from the rest point, the frequency decreases to a moderate level as Q shortens (Figure 5C), and the afterdepolarization disappears. As noted in Figure 2A, 1-bursts are relatively flat during their quiet spells Q. The spike amplitude is large and nearly constant over a wide range of frequencies.

In Figure 5, high frequency 1-bursts join a 'family of Ω-periodic solutions. At the meeting point the frequency is moderate. Moving into the family of Ω-periodics (Figure 5E), the frequency increases further. As noted in Figure 2B, activity steadily increases throughout the interspike interval. The Ω-periodic family terminates at very high frequency (Figures 5F and 2A). High frequency Ω-periodic solutions have significantly smaller spikes than do low frequency Ω-periodic solutions, by contrast with 1-bursts.

I.6. Fine Structure of Ω-Periodic Waves

Figure 6 illustrates some detailed predictions of the generalized Hodgkin-Huxley model. These predictions describe the covariation of several properties of Ω-periodic solutions that apply to all model cells, whether or not they admit bursts. If a change of initial data causes a higher frequency of Ω-periodic spikes, then it also causes a lower amplitude of spikes, a smaller post-spike hyperpolarization, a lower wave speed, and a lower spiking threshold. In Figure 2B, the frequency, amplitude, and post-spike hyperpolarization of snail yel-

low cell spikes all covary as predicted, thereby strengthening our
contention that Figure 2B illustrates a family of Ω-periodic solutions.

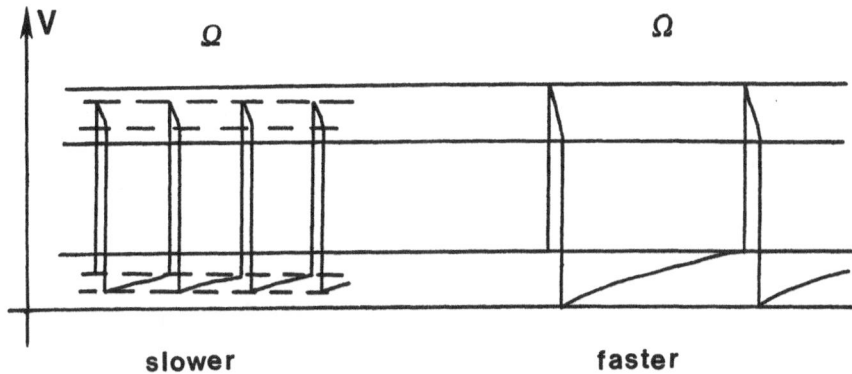

Figure 6:   Correlated properties of Ω-periodic solutions

higher frequency	lower frequency
lower amplitude	higher amplitude
smaller post-spike	larger post-spike
hyperpolarization	hyperpolarization
low speed	moderate or high speed
lower threshold	higher threshold

I.7. Effects of Drugs on Signal Patterns: The Inverse Problem

Another type of data that illustrate these predictions are drug
effects on nerve cell signal patterns. A drug may cause complex chemical reactions in many intracellular and intercellular subsystems.
Most if not all of these reactions cannot be directly observed. The
present approach suggests a new method to help classify which of
these reactions are due to intracellular changes, notably changes in
the parameters of a Hodgkin-Huxley mechanism, and to generate inferences about which intracellular parameters may have changed. The

method studies the observable parametric changes in a cell's electrical signal patterns and uses these changes to make inferences about corresponding changes in unobservable cell parameters. The attempt to infer underlying mechanisms from observable properties is often called an <u>inverse problem</u>.

Figure 7 depicts a drug effect on a giant neuron of the snail <u>Helix pomatia</u> (Lábos and Láng, 1978) that can be interpreted in terms of our parametric predictions about bursts and Ω-periodic waves. Figure 7A illustrates the regular periodic output of a giant neuron in an

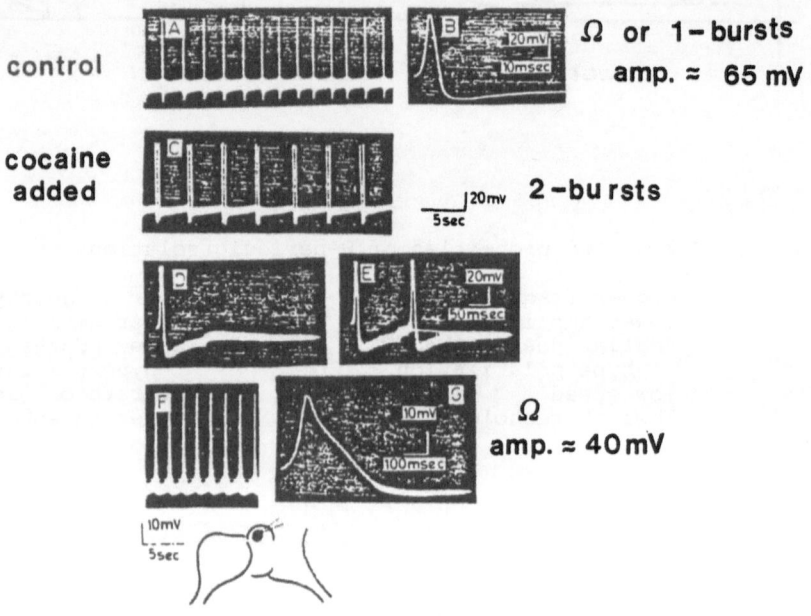

Figure 7: Recordings from snail (<u>Helix</u>) neuron. (A)-(B): control; (C)-(E): 10-25 minutes after administration of cocaine; (F)-(G): after 30 minutes (Lábos and Láng, 1978, p.179). Compare (A) with (F): in (A) the frequency is lower, the amplitude is higher, and the post-spike hyperpolarization is larger - all as predicted in Figure 6.

isolated subesophageal ganglion before cocaine was applied to the bathing fluid. This regular periodic solution is either a periodic 1-burst or an Ω-periodic solution. Recall from Figure 5 that periodic 1-bursts can merge with Ω-periodics in a single family of solutions, so no single picture can tell them apart.

During the first 30 minutes after the cocaine was applied, the regular periodic solution was transformed into a periodic 2-burst solution (Figure 7C). After the first 30 minutes, the effects of the drug started to wear off, leading to the regular periodic solution of Figure 7F, before this solution approached the original waveform in Figure 7A. Despite the incomplete nature of these data for solving the inverse problem, a comparison of Figures 7A and 7F is informative in the light of Figure 6. Figure 6 suggests that Figure 7F depicts an Ω-periodic solution, not a periodic 1-burst. Note that the amplitude of spikes is larger, the frequency of spikes is lower, and the post-spike hyperpolarization is greater in Figure 7A than in Figure 7F, just as predicted in Figure 6. To draw a more complete portrait of underlying cellular changes, a parametric series of dose-dependent waveforms at a regular succession of times after dose would be most helpful.

I.8. Parametric Structure of Bursts

Just as with 1-bursts and Ω-periodics, a series of predictions can be made about N-bursts. The family of all such N-bursts will be called HH bursts to distinguish them from other burst types, such as the parabolic bursts and paroxysmal bursts that occur in data and the FN bursts that are solutions of the FitzHugh-Nagumo model. These other burst types will be discussed later.

As shown in Figure 3A, an HH burst with many spikes per burst moves away from the rest point towards an Ω-periodic solution. Hence its spiking frequency speeds up and becomes more constant late in the

burst. As with l-bursts, the potential in such an HH burst is nearly
flat during the quiet spell Q. Figure 8A contains a periodic bursting
solution that illustrates these predictions. This record depicts spon-

A

Spontaneous action potentials in monkey
epileptic cortex (Atkinson and Ward, 1964,
p.291).

B

Bursts in the lobster stomatogastric ganglion (Russell and Hartline, 1978, p.454).

C

0.5 s   Bursts in the motor neuron controlling
expiration in the dragonfly (Mill, 1977,
p.193).

Figure 8:   Typical HH burst patterns in intracellular recordings. Note
the increasing frequency within the burst; the flat inter-
burst interval; and the afterdepolarization in (B).

taneous action potentials that were recorded from the monkey epileptic
cortex (Atkinson and Ward, 1964, p.291). Figure 8C depicts bursts of
similar form that were recorded from a motor neuron which controls
expiration in the dragonfly (Mill, 1977, p.193). In both figures, the

frequency of spikes increases and levels off before the burst suddenly terminates. Figure 8B depicts a bursting solution with so few spikes per burst that the frequency never speeds up, as also occurs in Figure 3B.

Another prediction for HH bursts is that the length of the quiet spell Q between bursts increases with the number of spikes in each burst, as illustrated in Figure 9 within snail yellow cells. Due to this property, there is a tendency for a fixed number of spikes to occur within a sufficiently long time interval whether these spikes are

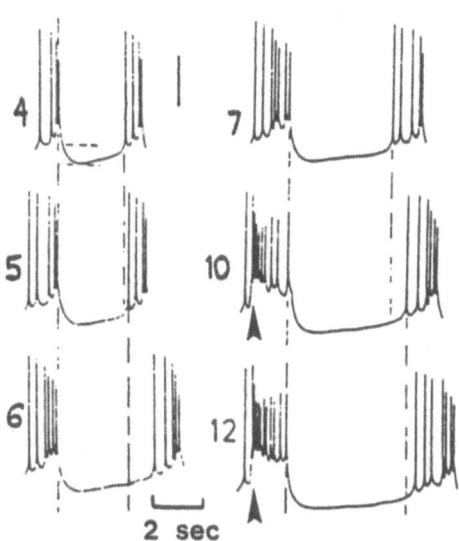

Figure 9: As predicted, the length of the quiet spell increases with the number of spikes in the previous burst in snail (<u>Lymnae stagnalis</u>) yellow cells (Benjamin, 1978, p.209).

grouped into 1-bursts, 2-bursts, and so on. This tendency is illustrated in Figures 7A and 7C. There the application of cocaine resulted in the clustering of spikes into 2-burst doublets, but the average frequency remained the same: each trace has 14 spikes within approximately 34 seconds.

Even in cells wherein the average spiking frequency remains
entirely invariant when the bursts change their structure, these changes can significantly alter the firing of the cells which receive the
bursts (Calvin, 1972). This is because a large number of rapidly
occurring spikes within a burst can drive the target cell potential to
a much higher asymptote than the same number of spikes spread more
thinly through time, as Figure 10 schematically illustrates. Carpenter
(1979, 1981) makes other predictions concerning the structure of HH
bursts, notably their stability under noisy perturbations.

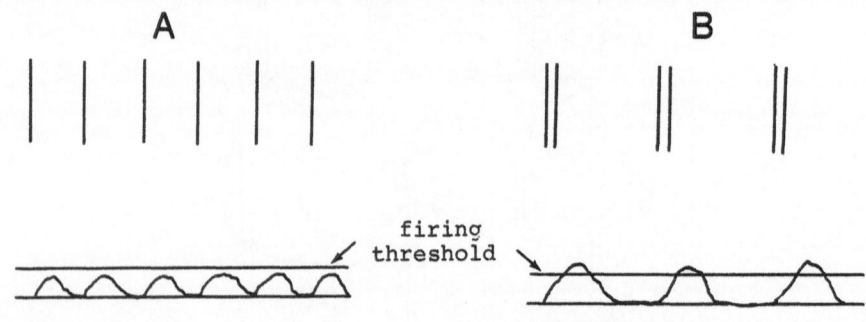

Figure 10:   Top row: pre-synaptic spike train. Bottom row: post synap-
             tic transmitter concentration. In (A) and (B), the average
             spike frequencies are identical. However, the cell which
             bursts (B) is more excitable than the cell (A) with a reg-
             ular spike pattern.

I.9. Finite Wave Trains and Chaotic Waveforms: Aperiodic Phenomena

In addition to the periodic bursts and Ω-periodic solutions, several classes of aperiodic waveforms that were not previously known to
be consequences of Hodgkin-Huxley dynamics have been mathematically
proved to exist (Carpenter, 1977a, 1977b, 1979, 1981).

The simplest aperiodic waveforms are the finite wave trains.
These solutions generalize the single spike in Figure 5A. They contain
a single burst, preceded and followed by subthreshold activity, whose
internal structure is akin to the bursts that occur within periodic
bursting solutions. It has been proved that the speed with which such
an N-burst travels down the axon is an increasing function of N.

More complex aperiodic waveforms can also exist. Under special
hypotheses, all possible sequences of bursts can be generated within a
single model neuron. In other words, some model neurons can support an
infinite dimensional temporal code despite the fact that they are defined by a five-dimensional dynamical system. More precisely, within
such a system, given any sequence $\{N_1, N_2, N_3, ...\}$ of positive integers, there is a solution with N_1 spikes in the first burst interval,
N_2 spikes in the second burst interval, N_3 spikes in the third burst
interval, and so on. Moreover, the solutions are ordered lexicographi-
cally by the wave speed ω in the following sense. Suppose that $\{N_1,
N_2, N_3, ...N_K...\}$ and $\{M_1, M_2, M_3, ...M_K...\}$ represent two such solu-
tions, with speed ω_N and ω_M, respectively. If we compare these sequences term-by-term, N_1 with M_1, N_2 with M_2, and so on, there always is
a first pair of terms that differ unless the sequences are identical.
Denote this first pair by (N_K, M_K). Then $\omega_N < \omega_M$ if and only if $N_K < M_K$.
More complex, even chaotic, waveforms are also possible, again under
special hypotheses but without adding any more variables to the model.

I.10. Parabolic and Paroxysmal Bursts: Augmented Hodgkin-Huxley Models

There exist bursting patterns in vivo that are not consequences
of the basic Hodgkin-Huxley dynamics with three ionic variables m, n,
and h. Some of these bursts can, however, be generated by Hodgkin-
Huxley models with four ionic variables. Figure 11A depicts a periodic
bursting pattern of this type, the so-called parabolic bursts that

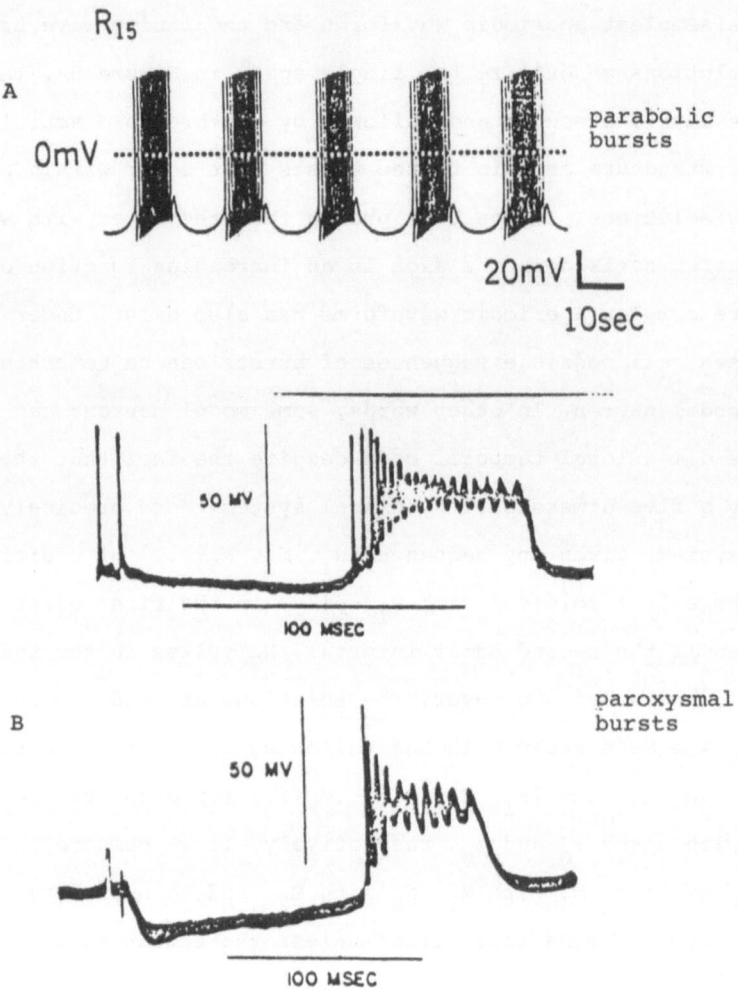

Figure 11: (A) Parabolic bursts in <u>Aplysia</u> abdominal ganglion(Roberge, et al, 1978, p.392).
(B) Paroxysmal bursts in the cat hippocampus (Kandel and Spencer, 1961, p.245).

have been studied in the <u>Aplysia</u> abdominal ganglion (Roberge, <u>et al</u>, 1978, p.392). Within a parabolic burst, the spiking frequency first increases, then decreases, before shutting off. The first part of the parabolic burst, wherein spiking frequency increases, resembles an HH burst (Figures 8A and 8C). Due to this fact, a parabolic burst form can be generated if there exists another slow process acting on the time scale of seconds, rather than the millisecond time scale of the Hodgkin-Huxley currents, that interacts with the three faster currents. Because the time scale of this additional process is relatively slow, it acts like a parametric change that moves the burst away from its Ω-periodic solution towards the rest point and thereby slows the frequency of spikes before the burst terminates (Carpenter, 1979). Either a slow accumulation of extracellular potassium or an additional slow potassium current can produce this effect (Faber and Klee, 1972).

Paroxysmal bursts, depicted in Figure 11B, occur in cell bodies rather than being propagated down axons (Kandel and Spencer, 1961). Such a burst pattern can formally be generated by antidromic (backwards) flow of potential from the cell body to the cell dendrites which, in turn, further depolarize the cell body and raise the baseline of burst activity to its plateau level.

A related type of burst which rides a plateau is illustrated in Figure 12. This periodic bursting solution is generated by periodic

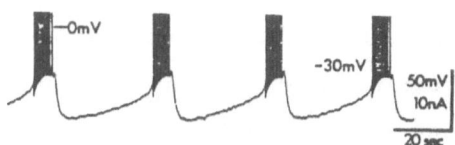

Figure 12:   A burst riding the crest of a slow potential wave (Otala lactea, cell 11) (Barker and Smith, 1978, p.380). The large oscillations in the baseline potential are not part of HH bursts, although small oscillations may be present.

pacemaker activity, or a slow depolarization shift, (DPS), on which spikes are superimposed when the potential is suprathreshold. Here the mechanisms of interest are those which generate the DPS, rather than the spikes _per se_. This is the type of burst discussed in Plant and Kim (1975).

I.11. FitzHugh-Nagumo Bursts

Burst solutions have recently been reported to occur in the Fitz-Hugh-Nagumo model (Evans, Fenichel, and Feroe, 1982; Feroe, 1982; Hastings, 1982). This model is similar to a generalized Hodgkin-Huxley model with two rather than three ionic processes. It was originally motivated by the observation that since potassium activation (n) tends to increase while sodium inactivation (h) tends to decrease, the sum n+h does not vary too much (FitzHugh, 1961; Nagumo, Arimoto, and Yoshizawa, 1962). Hence one degree of freedom was removed from the original Hodgkin-Huxley model by assuming that n+h is identically constant. Other simplifications in the model were made to represent it as a tunnel diode.

In vivo, the processes of sodium inactivation (h) and potassium activation (n) are relatively _slow_. In the Hodgkin-Huxley model, this fact becomes the mathematical hypothesis that the model's eigenvalues are real numbers. A necessary requirement for the Evans _et al_ (1982) and Hastings (1982) analysis is that the corresponding process in the FN model be _fast_ in order to create eigenvalues that are complex numbers. In the original HH model, n and h could only change this quickly if the model's temperature variable were set at twice the normal temperature (Centigrade).

Independent of discussions about the physical plausibility of the FN-burst hypothesis, one can test for FN bursts and HH bursts _in vivo_ by their parametric properties. FN bursts are obtained by perturbing off a single spike, so the spikes within such a burst are evenly and widely spaced, by contrast with the typical speed-up of HH bursts depicted in Figures 8A and 8C.

II. THE TRANSDUCTION OF LIGHT BY VERTEBRATE PHOTORECEPTORS

II.1. The Turtle Cone and the Dynamics of Chemical Transmitter Substances

In this section we consider the electrical response of cells in another model system, the turtle cone. The discussion will have several points in common with the previous section on single nerve cell dynamics. The turtle cone, like the squid axon, is important not just in itself, but because it is a convenient experimental preparation in which to investigate a general neural phenomenon: the transduction of light into an electrical response. Many experiments have indicated that this transduction process is mediated by a chemical transmitter within the photoreceptor. Our discussion will compare two models for the interactions of light, transmitter, and photoreceptor potential: the unblocking model of Baylor, Hodgkin, and Lamb (BHL) (1974b) and our own gating model (Carpenter and Grossberg, 1981).

As with the Hodgkin and Huxley work on the squid giant axon, the generative work on turtle cones includes an exhaustive series of parametric experiments (Baylor and Hodgkin, 1973, 1974; Baylor, Hodgkin, and Lamb, 1974a, 1974b) that led to the unblocking model. Both the unblocking model and the Hodgkin-Huxley model begin with the basic membrane equation, which is then augmented by auxiliary equations. Our

analysis of turtle cone dynamics will, however, be very different from
the analysis in Section I. In their article, Baylor, Hodgkin, and Lamb
(1974b) do an extensive numerical analysis of their model, and note
that, despite its complexity, it fails to meet important data. Their
model's difficulties seem to center on how the chemical transmitter
mediates between light and potential. We realized that an alternative
model of transmitter dynamics (Grossberg, 1968, 1969) could provide
intuitively simpler and better quantitative fits to the turtle cone
data. This gating model has additional appeal to us because it was
derived from a general principle of neural design and has been helpful
in explaining transmitter-mediated data in a variety of other neural
systems (Grossberg, 1975, 1980, 1982a,b).

Thus in this section, our discussion will approach the problem of
how to choose between two models of a complex biological phenomenon.
This task is rendered all the more difficult by the fact that the
light transduction process involves multiple stages that are difficult
to experimentally disentangle. No model can hope to include all the
interactions that will eventually be disclosed by ever finer physiological, biochemical, and molecular techniques.

We approach this task by offering parametric predictions that can
characterize whether the type of transmitter gating process that we
have in mind is rate-limiting in any body of data, whether in a photoreceptor or not. Thus our approach is to explicate the parametric implications of a basic neural design no less than to try to understand
a particular body of data. In the case of vertebrate photoreceptors,
experimental articles based on electrophysiological recording typically interpreted the major electrical effects to occur in the outer segment of the photoreceptor. More recent experiments using suction elec-
trodes that can electrically isolate the photoreceptor outer segment
from its inner segment suggest that these electrical effects are not

necessarily localized at the outer segment (Nunn, Matthews, and Baylor, 1980; Yau, McNaughton, and Hodgkin, 1981). Our analysis does not
attempt to specify the anatomical locations at which a transmitter or
transmitters may act, but rather provides parametric evidence for
whether a rate-limiting transmitter gating step occurs at some stage
in the generation of photoreceptor potential.

II.2. The Unblocking Model

Figure 13 illustrates a schematic version of the first stages of
vertebrate photoreceptor dynamics, as represented by BHL. Light enters

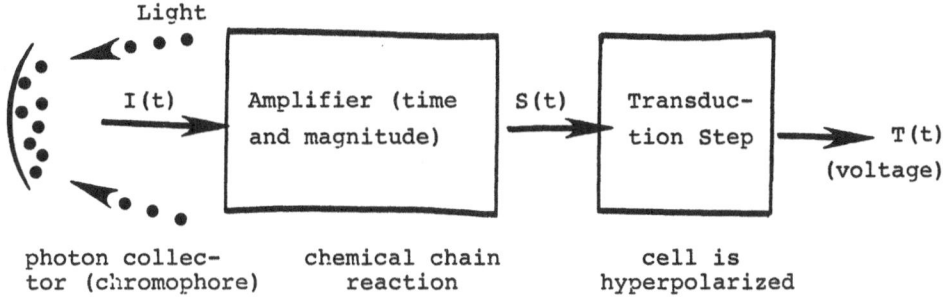

Figure 13:   Schematic view of the early stages of light processing in
            the vertebrate photoreceptor.

the front of the eye, and photons are collected at the back of the eye,
in the chromophore. The light signal, $I(t)$, is amplified, both in time
and magnitude, by a chemical chain reaction. Such an amplification step

enables the eye to register even 1 or 2 photons by generating enough activity to translate these microscopic signals into a voltage change which can be transmitted to the brain. The output of the chain reaction is $S(t)$. Next comes the transduction transformation

$$S(t) \rightarrow T(t) \tag{14}$$

whereby the photoreceptor is hyperpolarized by light and the electrical signal $T(t)$ is generated. The signal $T(t)$ influences the activity of subsequent retinal layers and the brain. We will focus our attention on the transduction transformation (14).

The BHL model of the transduction transformation is summarized below.

Membrane Equation:

$$\tau_L \frac{dV}{dt} = E - V(1 + G_f + G_i) \tag{15}$$

Light-Sensitive Ionic Conductance:

$$G_i = \frac{\bar{G}_i}{1 + \frac{z_1}{K}} \tag{16}$$

Time-Average of Prior Voltage Transform:

$$\tau_f \frac{dG_f}{dt} = F(V) - G_f \tag{17}$$

Logistic Transform of Voltage:

$$F(V) = \frac{\bar{G}_f}{1 + \exp\left[\frac{(V - V_f)}{V_\varepsilon}\right]} \tag{18}$$

Linear Chain Reaction to Light:

$$\left(\frac{d}{dt} + \alpha\right)^{n-1} y_{n-1} = \alpha^{n-2} I(t) \tag{19}$$

Blocking Substance:

$$\frac{dz_1}{dt} = \alpha y_{n-1} - \kappa_{12} z_1 + \kappa_{21} z_2 \tag{20}$$

Unblocking Substances:

$$\frac{dz_2}{dt} = \kappa_{12} z_1 - (\kappa_{21} + \kappa_{23}) z_2 + \kappa_{32} z_3 \tag{21}$$

$$\frac{dz_3}{dt} = \kappa_{23} z_2 - (\kappa_{32} + \kappa_{34}) z_3 \tag{22}$$

Gain Control:

$$A\kappa_{21} = \kappa_{12} = \bar{\kappa}_{12} + \nu z_2 \left[\frac{\kappa_{12M} - \bar{\kappa}_{12}}{\kappa_{12M} - \bar{\kappa}_{12} + \nu z_2} \right]. \tag{23}$$

Just as in the HH model, $V(t)$ in (15) is the transmembrane voltage. In the BHL model, however, $V(t)$ is assumed to be spatially homogeneous; possible spatial interactions of different photoreceptor segments are not considered. Function G_i in (15) and (16) is a light-sensitive ionic conductance. In (16), G_i is a decreasing function of z_1 which represents the concentration of a blocking substance that is hypothesized in (20) to be the output of the light-initiated chain reaction

$$I(t) \rightarrow y_1 \rightarrow y_2 \rightarrow y_3 \rightarrow \cdots \rightarrow y_{n-1} \rightarrow z_1 \tag{24}$$

that is defined by (19). According to equations (15), (16), (19) and (20), light increases the amount of blocking substance and thereupon hyperpolarizes the membrane by decreasing G_i in (15).

The blocking substance z_1 is removed by the unblocking substances z_2 and z_3 according to equations (20)-(23). At first glance, the blocking-unblocking equations (20)-(22) appear to be linear. However the z_2 terms in the definitions of κ_{12} and κ_{21} in (23) show that equations (20)-(22) are highly nonlinear. The blocking-unblocking hypothesis is the key component of the BHL model.

The conductance G_f in (15) also depends nonlinearly on the voltage V via equations (17) and (18). Its significance in the BHL model will be reviewed when we discuss relevant data in Section II.5.

II.3. The Gating Model: Unbiased Transmitter-Mediated Signalling

The gating model of the transduction transformation can be derived as an answer to the following question:

What is the simplest law whereby one nerve cell site can send unbiased signals to another nerve cell site?
If S(t) is the input to one cell site and T(t) is the output to the next cell site, then a linear relationship such as

$$T = SB \qquad (25)$$

is clearly the simplest law for unbiased transmission, where B is a positive constant. Here the outgoing signal T is directly proportional to the incoming signal S, so the signal is relayed perfectly.

This is not the end of the discussion when the output signal T(t) is due to the release of a chemical transmitter substance z(t) in response to the input signal S(t). Then we must face the issue of how a large and sustained signal S(t) is prevented from depleting z(t) and thereby causing a progressively smaller signal T(t).

From this perspective, equation (25) may be replaced by the pair of equations

$$T = Sz \tag{26}$$

and

$$z \cong B. \tag{27}$$

Equation (26) says that transmitter z is released at a rate (proportional to) T in response to signal S. In other words, z gates S to
generate T, or T is caused by a mass action interaction between S and
z. By (26), more transmitter is released if either the incoming signal
S or the amount of available transmitter z is increased.

Equation (27) simply requires that the amount of transmitter seek
a constant level B, as in (25), so that the sensitivity of T to S will
not decrease due to transmitter loss. If equations (26) and (27) can
be simultaneously implemented, then the perfect transmission described
by equation (25) will be assured. The gating model is derived from
hypotheses that aim to maintain unbiased transmission when the cell
sites in question signal each other via a depletable chemical transmitter. Herein lies the intuitive appeal and generality of the gating
concept.

The simplest dynamical equation that is capable of simultaneously
summarizing equation (26) and (27) is the following:

Transmitter Accumulation-Depletion Equation

$$\frac{dz}{dt} = A(B-z) - Sz , \tag{28}$$

where A,B > 0. In (28), term A(B-z) says that z accumulates until it
reaches the target level B, as required by (27). Term -Sz says that
transmitter is depleted at the rate T, as required by (26).

II.4. <u>Unbiased Transmission by Miniaturized Cells: Light-Induced</u>
<u>Enzymatic Activation of Transmitter Accumulation</u>

Equation (28) is the simplest dynamical law that might possibly subserve unbiased chemical transmission between cell sites, but it is not adequate in cells that possess certain properties shared by photoreceptors. This is because the rate of transmitter accumulation (A) in term $A(B-z)$ may be small compared to the rate of transmitter depletion (S) in term $-Sz$ if the signal S can become large, as in a photoreceptor that can respond to a wide dynamical range of light intensities $I(t)$. Then $z(t)$ can become depleted significantly below its asymptotic level B, and loss of sensitivity in T's response to S can occur. Thus equation (28) alone does not solve the following.

Problem: A large input signal $S(t)$ can deplete $z(t)$ and cause <u>habituation</u> or desensitization of $T(t)$.

How can a cell maintain its sensitivity in spite of large fluctuations in light input intensity? Two types of solution can be contemplated in (28). The first solution, which we do not adopt, leads to the type of reaction to light that is schematized by the BHL chain reaction (24) leading from light input I to transmitter reaction z_1.

If a large storage depot of transmitter is maintained, then even large signals S will deplete only a small fraction of the depot. Hence, if $B \gg 1$, then even large (but bounded) signals S will cause only a small reduction in the ratio zB^{-1}.

If a large storage depot of transmitter existed within each photoreceptor, then this depot would enlarge each photoreceptor's volume, and would thereby reduce the number of photoreceptors that could be packed into a fixed retinal area. Improving the dynamical sensitivity of each photoreceptor using this method would reduce the spatial sensitivity of the retina as a whole.

We can therefore reformulate our design problem as follows.

Problem: How can a spatially miniaturized photoreceptor maintain its
sensitivity to large fluctuations in input intensity?

If in equation (28), B is not large relative to S, then an alternative solution is to let the rate of transmitter accumulation (A) increase as a function of S to counterbalance the rate S of transmitter
depletion. Otherwise expressed, we suppose that there exists a light-
induced enzymatic activation of the transmitter accumulation rate.
According to this hypothesis, the effects of light on the photoreceptor do not merely follow a serial chain of transduction steps. Rather,
light activates parallel pathways which later mutually interact to
generate the cone potential.

The simplest mass action law for a light-activated accumulation
rate is

$$\frac{dA}{dt} = -C(A-A_o) + D[E-(A-A_o)]S. \tag{29}$$

Here, A_o is the baseline level of $A(t)$ in the dark ($S=0$). An incoming
light-induced signal S tends to drive the accumulation rate $A(t)$ towards its maximal level A_o+E.

We have found that the two equations (28) and (29) numerically fit
the most difficult BHL data better than the full BHL model (15)-(23).
Indeed a single equation derived from (28) and (29) fits these data
better than the full BHL model. This is true even if these equations
are coupled to the cone potential $V(t)$ in the simplest possible way by
making the change in T proportional to the change in V. Wherever such
a coupling or more realistic couplings occur, whether in the outer segment or the inner segment, does not influence the meaning of the gating
step or the goodness of numerical fit.

The simpler gating model assumes that enzymatic activation occurs
quickly relative to z's equilibration rate. Then A in (29) is always

approximately at its equilibrium level

$$A(S) = A_o + \frac{EGS}{1+GS} \tag{30}$$

where $G \equiv DC^{-1}$. This approximation leads to

Gating Model I

$$T = Sz \tag{26}$$

$$\frac{dz}{dt} = A(S)(B-z)-Sz \tag{31}$$

where $A(S)$ is defined by (30). When the enzymatic activation of A proceeds at a finite rate relative to z's equilibration rate, we study the

Gating Model II

$$T = Sz \tag{26}$$

$$\frac{dz}{dt} = A(B-z)-Sz \tag{28}$$

and

$$\frac{dA}{dt} = C(1+GS)[A(S)-A], \tag{29}$$

where $A(S)$ is defined by (30).

II.5. Double Flash Experiments

An important difference between the chain reaction of the BHL model

$$I \rightarrow y_1 \rightarrow y_2 \rightarrow \cdots \rightarrow y_{n-1} \rightarrow z_1 \rightarrow \tag{24}$$

and a chain reaction of a gating model

$$I \rightarrow y_1 \rightarrow y_2 \rightarrow \cdots \rightarrow y_{n-1} = S \rightarrow Sz \rightarrow \qquad (32)$$

is illustrated by a double flash experiment (Figure 14). Baylor, Hodgkin, and Lamb (1974a) found that a bright flash causes the potential

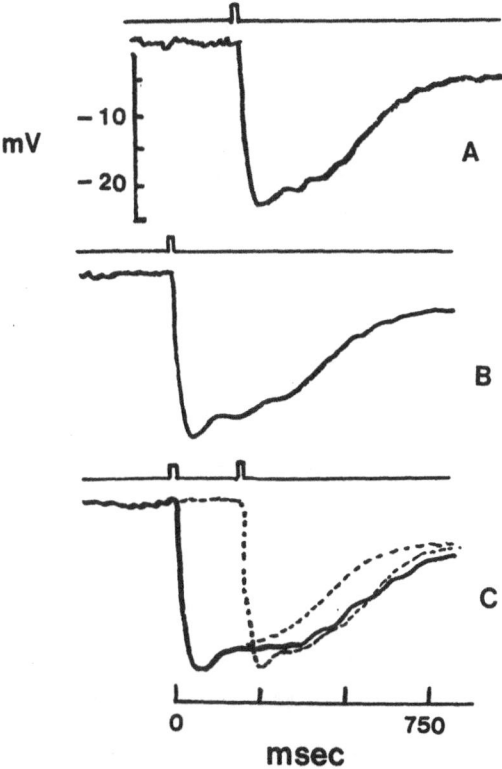

Figure 14: Effect of a bright conditioning flash on the response to a
subsequent bright test flash. (A) Response to test flash
alone. (B) Response to conditioning flash alone. (C) Response to both flashes, with the upper two responses dotted
in. Redrawn from Figure 15 (Baylor, Hodgkin, and Lamb,
1974a, p.716).

to overshoot before settling to a plateau level that is maintained for a while before the potential returns to equilibrium. A second bright flash that occurs while the potential is at the plateau value caused by the first flash does not cause an overshoot even though it does extend the duration of the chain reaction.

Given a chain reaction like (24), it is not possible to understand how the duration of the chain reaction can be prolonged without increasing the concentration of internal transmitter (Detwiler, Hodgkin, and McNaughton, 1980, pp.222-223). Since, however, the voltage does not change in response to this change in transmitter concentration, the voltage must be saturated, or insensitive to, the transmitter increment. In the BHL model, the definition of the conductance G_f in equations (15), (17), and (18) is made to guarantee the voltage's insensitivity to a second flash.

Notwithstanding the possible physical truth of these assertions, the gating model provides an alternative explanation: The transmitter term z in the chain reaction (32) can equilibrate via equation (28) or (31) to the first flash, thereby causing an overshoot in potential, and can thereafter remain insensitive to a second flash that occurs while S remains at its plateau value, without preventing the new flash from prolonging the duration of the chain reaction. See Carpenter and Grossberg (1981, pp.15-16) for further details. This is perhaps the critical difference between an unblocking model and a gating model. A more critical test between the models might be made using a parametric series of double flash experiments in which the second flash is twice as intense as the first.

The remainder of this review of photoreceptor models will compare how the unblocking and gating models fit a demanding body of parametric data, how the gating step can be coupled to a membrane equation, and how the gating model can be parametrically tested by blocking the light-induced enzymatic activation step.

II.6. <u>Parametric Studies of Flashes on Backgrounds: Differential</u>

    <u>Reactions in the Energy and Time Domains</u>

    A critical series of parametric experiments which generated para-

doxical data was reported in Baylor and Hodgkin (1974). The Gating

Model I with just one dynamical equation fits these data better than

the full BHL model. The Gating Model II with two dynamical equations

provides a significantly better fit. In these experiments, (Figure

15), a brief (11 msec.) flash of fixed size (δ) is superimposed on a

constant level of background illumination (I_o). The cone potential is

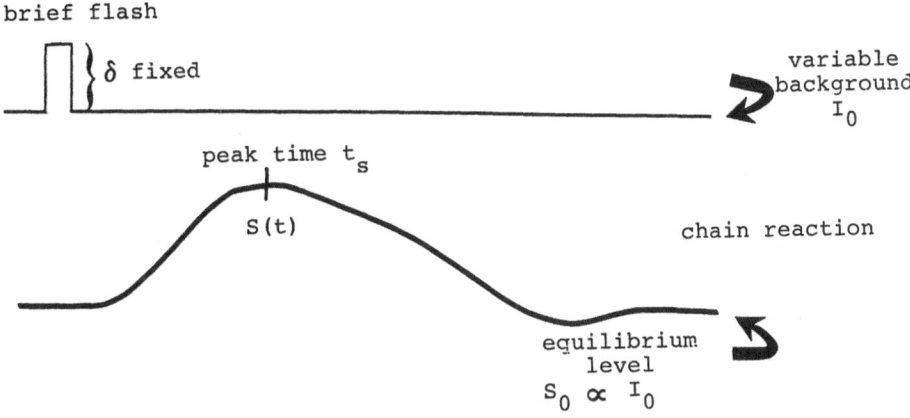

Figure 15: This experiment, from (Baylor and Hodgkin, 1974), measures
the photoreceptor's response to a brief flash of fixed
size, superimposed on a level of background light which ranges from dark to very bright.

allowed to equilibrate to the background level I_o before the flash

occurs at time t=0. As I_o is parametrically increased across several

log units, the cone's voltage responses to the fixed flash are measured for t ≥ 0. The results of these experiments are plotted in Figure
16a. Each curve in Figure 16a represents the amount of hyperpolarization caused by the flash δ relative to the baseline level of hyperpolarization caused by the background level I_o.

As I_o is increased, the peak hyperpolarization decreases, as is
intuitively plausible. However, the <u>times</u> at which these peak responses occur first decrease as I_o increases, but then, at sufficiently
large values of I_o, begin to increase. Thus a monotonic change in the
energy domain coexists with a non-monotonic change in the time domain.
Such a differential parsing of energy and time also occurs in the
double flash experiments (Section II.5) wherein a second flash influences temporal but not energetic measures, by contrast with a first
flash.

Numerical solutions of the BHL model are shown in Figure 16b.
Note that the scales in the lower two graphs differ from those in Figure 16a, and that the peak sizes are off by a factor of 10 in the lowest graph. Also the turnaround occurs at too low a light intensity,
and the time of the turnaround is too late.

Results for the Gating Model I are given in Figure 16c. Here, the
peak sizes at large I_o are also off by a factor of 10, but the time
and intensity of the turnaround are closer to the data. Recall that
Model I consists of a single linear differential equation.

Results for the Gating Model II, in Figure 16d, fit the data
well: peak size, turnaround time, and light intensity at the turnaround are all close to those in Figure 16a. The times at which peak
response occurs, as functions of log I_o, are plotted in Figure 17 for
the Baylor-Hodgkin (1974) data, the BHL model, and the two gating
models.

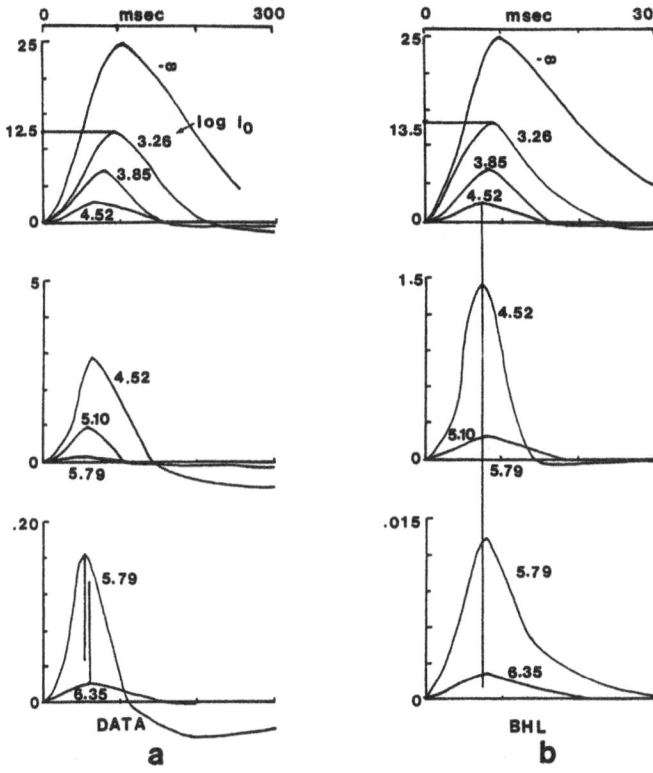

Figure 16: Intracellular response curves x(t)-x_O showing the effect of a flash superimposed on a background light of fixed intensity, I_O. Each horizontal axis represents the time, in msec., since the middle of the 11 msec. flash. Each vertical axis is scaled so that the peak value of x(t)-x_O=x(t) in the dark is equal to 25. The number beside each curve equals $\log_{10}(I_O)$, where I is calibrated so that when $\log_{10}I_O$ = 3.26, the peak of x(t)-x_O equals 12.5. (a) The Baylor-Hodgkin (1974) data. (b) The Baylor-Hodgkin-Lamb model predictions redrawn from (Baylor, Hodgkin, and Lamb, 1974b, p.785).

144

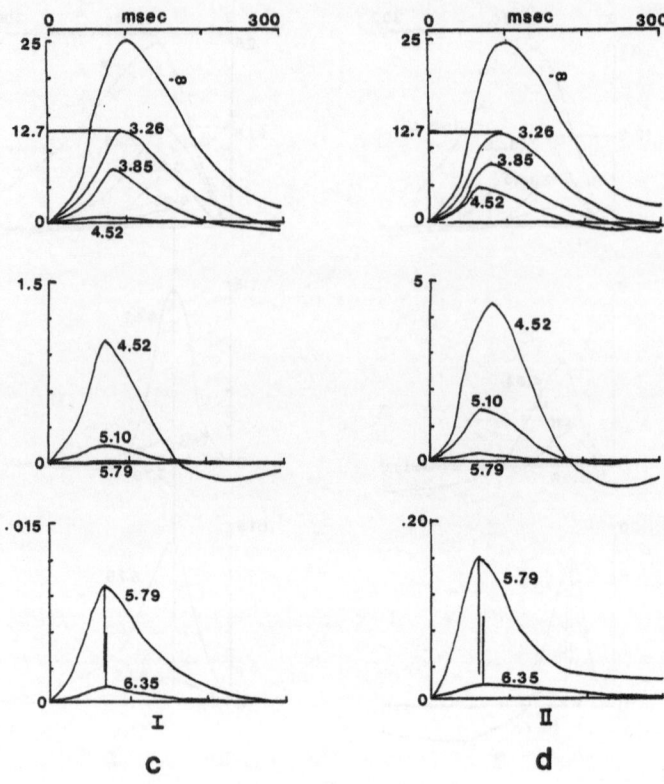

Figure 16 (cont.): (c) Predicted response of Gating Model I (Equation
(9)). (d) Predicted response of Gating Model II
(Equations (7)-(8)). For each of Models I and II,
x(t) is proportional to S(t)z(t). Note that the
vertical scales of (a) and (d) are the same, and
the vertical scales of (b) and (c) are the same.

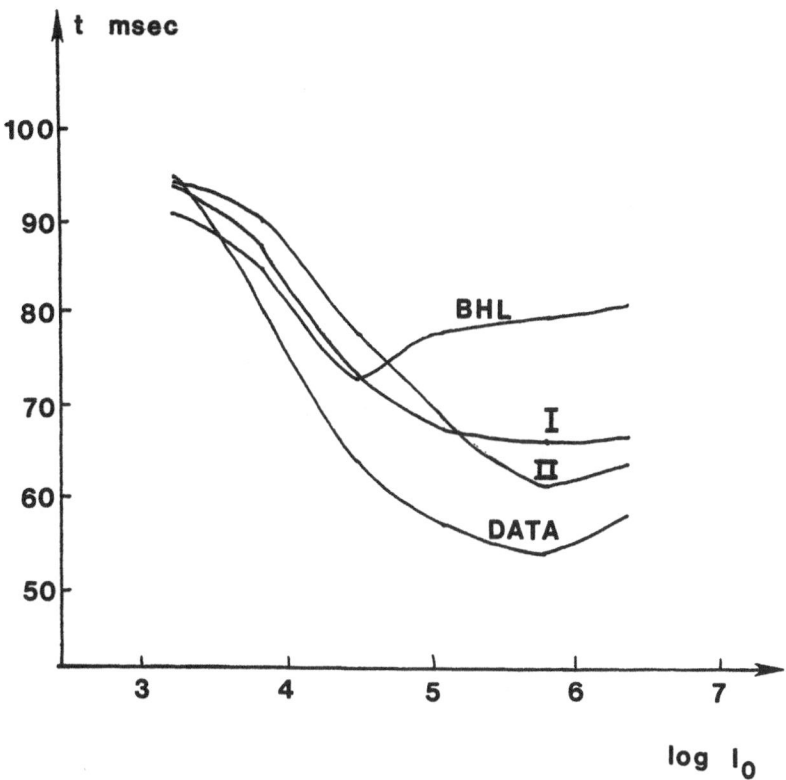

Figure 17: Times at which the peak hyperpolarizations occur for the Baylor-Hodgkin (1974) data and the three models. Note that the input intensity at which the turnaround occurs and the dynamic range of peak times are much too small in the BHL model. Baylor, Hodgkin, and Lamb (1974b) considered this the most serious defect of their model.

II.7. Locus of the Transmitter Gating Stage

The model thus far has described only chain reaction and transmitter gating effects in keeping with our hypothesis that the gate is the rate-limiting cause of phenomena like the turnaround of potential peaks. Both conceptual completeness and the explanation of various other phenomena require that the gating step be coupled to the cone potential. For example, Baylor, Hodgkin, and Lamb (1974a) found during double flash experiments that an extra slow conductance accompanies the overshoot to the first flash but not the prolonged response, without overshoot, to the second flash. In the unblocking model, this property requires the hypothesis that an extra membrane channel exists with properties that control the slow conductance term. In a gating model, a slow conductance effect is found without hypothesizing an extra membrane channel (Carpenter and Grossberg, 1981, p.22). The slow conductance and its relationship to the overshoot of potential follow directly from the coupling of the gate to the potential.

Such properties of the gating model are invariant under model variations which place the coupling between gate and potential at different stages in the light transduction process. This fact is important to realize because electrophysiological data leave uncertain the exact stage at which a particular biochemical process occurs. This uncertainty has recently been reduced by the use of suction electrode methods that permit outer segment recordings to be made in isolation from inner segment electrical signals (Baylor, Lamb, and Yau, 1979). These methods suggest that various processes which produce overshoot phenomena occur in the inner segment, rather than the outer segment, as had previously been thought (Baylor, Lamb, and Yau, 1979; Nunn, Matthews, and Baylor, 1980). This modification in the locus of overshoot-related phenomena is compatible with the existence of a rate-limiting gating step, but indicates that such a step occurs later in the transduction process than previous data suggested.

II.8. Coupling the Transmitter Gate to Cell Potential

A gating signal can, in principle, either depolarize or hyperpolarize a cell's potential. Although this distinction is of great physical importance, many formal properties are the same in both cases. For example, $S(t)$ can be thought of either as the output of the chain reaction or as the effect of the chain reaction on the outer segment potential. Whichever interpretation of $S(t)$ is used, the gated signal $T(t)$ may then, in principle, either depolarize or hyperpolarize the cone potential. Whatever interpretation is needed to treat a particu-

lar case, the steady-state amount of hyperpolarization or of depolarization in response to a gated signal T=Sz is of the form

$$\frac{MT}{N+T} \ . \tag{33}$$

The rate with which the potential approaches asymptote (33) will depend on whether the cell is hyperpolarizing or depolarizing. However,
in cases where the potential reacts quickly relative to the gate's
reaction rate, as we assert occurs in turtle cones, the potential
change maintains the approximate equilibrium (33) whether the cell is
hyperpolarized or depolarized. In this case, the time and the background intensity at which the potential, via (33), experiences a turnaround in potential peak is approximately the same as the time and the
background intensity at which the gated signal T itself experiences
its turnaround. In fact, (33) is approximately proportional to T
except if it begins to saturate at values of T $\gg$ N.

The gating model is coupled to the membrane equation as follows.
We start with the spatially uniform version of the membrane equation
for the voltage V(t):

$$C_0 \frac{dV}{dt} = (V^+ - V)g^+ + (V^- - V)g^- + (V^P - V)g^P \ , \tag{34}$$

where V^+, V^-, and V^P are the excitatory, inhibitory, and passive saturation points; and g^+, g^-, and g^P are the corresponding conductances.

To couple the excitatory conductance g^+ to the gated signal T=Sz,
we use a simple mass action law. Below we describe the case wherein an
increase in T decreases g^+, and thereby hyperpolarizes V. Similar qualitative results are obtained if T increases rather than decreases g^+.
One difference between the two couplings is that a hyperpolarizing
coupling decreases V's reaction rate, whereas a depolarizing coupling

increases V's reaction rate.

The simplest mass action hyperpolarizing action is defined by

$$\frac{dg^+}{dt} = H(g_0 - g^+) - Jg^+T ,$$ (35)

where g_0 is the maximal number of open pores in the dark (T=0). Equation (35) says that closed pores, which number g_0-g^+, open at a rate H; and that the gated signal T closes open pores, which number g^+, at a rate J. Equation (35) can be rewritten in the form

$$\frac{dg^+}{dt} = H(1 + KT)\left[\frac{g_0}{1+KT} - g^+\right]$$ (36)

where $K = JH^{-1}$. If pores g^+ close quickly relative to the reaction rate of the gate z to light, then (36) implies that

$$g^+(t) \cong \frac{g_0}{1+KT} .$$ (37)

Assuming that pores close quickly, as in (37); that $g^-(t)$ is constant, say $g^-(t) \equiv g_1$; and setting $C_0 = 1$ and $g^P = 0$ for simplicity, the membrane equation (34) becomes

$$\frac{dV}{dt} = (V^+ - V)\frac{g_0}{1+KT} - (V - V^-)g_1 .$$ (38)

The equilibrium potential V_0 can be found by setting T=0 and $\frac{dV}{dt} = 0$ in (38). It is

$$V_0 = \frac{V^+g_0 + V^-g_1}{g_0 + g_1} .$$ (39)

The amount of hyperpolarization

$$x = V_0 - V \tag{40}$$

then obeys the equation

$$\frac{dx}{dt} = -\left(g_1 + \frac{g_0}{1+KT}\right)x + \frac{LT}{1+KT} , \tag{41}$$

where

$$L = Kg_1(V_0 - V^-). \tag{42}$$

Equation (41) can, in turn, be written in the form

$$\frac{dx}{dt} = \left(g_1 + \frac{g_0}{1+KT}\right)\left(\frac{MT}{N+T} - x\right) , \tag{43}$$

where

$$M = V_0 - V^- \tag{44}$$

and

$$N = \frac{g_0 + g_1}{g_1 K} . \tag{45}$$

If $x(t)$ reacts quickly relative to the reaction rate of $z(t)$, then
(43) implies

$$x(t) \cong \frac{MT(t)}{N+T(t)} , \tag{46}$$

as in (33).

When equation (46) is written in the form

$$\frac{Nx}{M-x} = T , \tag{47}$$

it provides a basis for comparison with the Baylor, Hodgkin, Lamb (1974b) equation

$$\frac{aU}{U_L - U} = \frac{z_1}{K} \tag{48}$$

relating amount of hyperpolarization U (in their notation) to the blocking variable z_1. In the gating model, the gated signal T = Sz replaces z_1.

II.9. Tests of Enzymatic Activation

Another point of comparison between the gating model and the unblocking model arises by considering the steady-state reactions of both models to a series of background lights. In the gating model, equations (26), (28), (29), and (30) imply a dependence of the form

$$T = \frac{PS(1+QS)}{1+RS+US^2} \tag{49}$$

where coefficient U is small compared to QR (Carpenter and Grossberg, 1981, p.12, p.21). In the unblocking model, the dependence is of the form

$$\frac{z_1}{K} = \frac{PS(1+QS)}{1+RS} \, . \tag{50}$$

Again the gated signal T in (49) replaces the blocking variable z_1 in (50). Equation (49) also contains the quadratic term US^2. No such term appears in the unblocking equation (50), which shows that this equation must break down at large values of S, since the amount of blocking substance must be bounded, whereas the right-hand side of (50) grows without bound as S increases.

In the gating model, the quadratic terms QS and US^2 are due to

the light-induced enzymatic reaction (29). If the enzymatic activation could be chemically inhibited, then the steady-state gating equation (49) would reduce to the equation

$$T = \frac{PS}{1+RS} \, ,$$
(51)

wherein no quadratic terms occur. Simultaneously, the turnaround of
potential peaks due to light flashes on parametric increases of background light would cease to occur; only a monotonic decrease in the
time to peak should be observed. Finally, suppose that the steady-
state potential is plotted against the logarithm of light intensity
after adapting the cone to a series of prior background intensities.
Then the cone's potential at a given log intensity is shifted as a
function of background intensity. Inhibition of the enzymatic step
should reduce the size of the shift due to high intensity lights by a
predictable amount (Carpenter and Grossberg, 1981, pp.24-25). These
and related parametric predictions can be used to test for the existence of a rate-limiting enzymatically modulated transmitter gating
step in photoreceptor and related cellular preparations.

III. <u>CIRCADIAN RHYTHMS AND OTHER BIOLOGICAL CLOCKS</u>

III.1. <u>A Chemical Gating Model of Circadian Rhythms and Motivational Cycles</u>

Because the gating concept can be derived from general principles about unbiased chemical transmission between cells (Section II.3), it should not be surprising that chemical gates may be found in a wide variety of neural systems. Whenever such a general principle can be identified, a classification of the properties whereby the principle manifests itself in data is desirable. Our discussion of photoreceptor dynamics has reviewed a type of data which exhibits parametric properties akin to those of an enzymatically activated rate-limiting transmitter gating step in a feedforward chain reaction. In such a reaction chain, one does not expect to find spontaneous large amplitude oscillations or sustained oscillations in response to brief inputs. When chemical gates are placed in feedback anatomies, by contrast, a wide variety of oscillations can occur.

A role for gates in feedback anatomies first arose in models of motivated behavior (Grossberg, 1972a, 1972b, 1975, 1980, 1982a, 1982b, 1982c). The model of circadian rhythms that we introduce here for the first time shares many formal similarities with these earlier models of gated feedback networks. These similarities clarify a sense in which motivated behaviors are controlled by a hierarchy of similar mechanisms with a circadian pacemaker at the foundation of the hierarchy, and permit us to understand how rhythmic properties arise in the motivated behaviors themselves (Moore-Ede, Sulzman, and Fuller, 1982, p.186). Due to the role of slow chemical gates in generating these rhythms, our model allows us to understand why certain chemical transmitters can oscillate according to a circadian rhythm, not only because they are driven by pacemaker output, but also because they form an integral part of the rhythm-generating mechanism (Binkley, 1979; Deguchi, 1979; Jouvet, 1974; Lewy, Wehr, Goodwin, Newsome, and Markey, 1980; Markowitz, Rotkin, and Rosen, 1981; Nestler, Zatz, and Greengard, 1982; Passouant and Oswald, 1979; Takahashi, Hamm, and Menaker, 1980; Tapp and Holloway, 1981; Wehr and Wirz-Justice, 1982; Zatz and Brownstein, 1979).

Three different biological processes, properly juxtaposed, are necessary to define a gated pacemaker: slow chemical gates, feedback, and competition. If any one of these processes is removed, a gated pacemaker ceases to oscillate. Using these ingredients, we have defined a pharmacologically predictive model of a circadian pacemaker. In their 1978 article, Daan and Berde note the essential weakness of previous models: "Most models of circadian oscillators have been abstract, in the sense that they include parameters not definable in terms of concrete physiological or biochemical processes...This feature makes them difficult to test and limits their heuristic or predictive value"(p.299). Because our model is physiologically grounded, it enables us to provide a unified explanation of circadian properties that have not previously been dynamically explained, such as period doubling, bimodal activity patterns, rhythm splitting, long-term after-effects, Aschoff's rule and its exceptions, characteristic phase leads and lags, seasonal modulation of activity, and differences between nocturnal and diurnal animals (Aschoff, 1960, 1979; Enright, 1980; Jouvet, Mouret, Chouvet, and Siffre, 1974; Moore-Ede, Sulzman, and Fuller, 1982; Pittendrigh, 1960, 1974; Wever, 1979; Winfree, 1980). For example,concerning long-term after-effects, Pittendrigh (1974) wrote in an important review article: "They are more widespread than the current literature suggests; they are not accounted for by any of the several mathematical models so far published; and they must be reckoned with in the mechanism of entrainment"(p.441). Our work pays particular attention to after-effects and provides examples of all the after-effect phenomena that Pittendrigh (1974) discusses.

III.2. Some Alternative Circadian Models

As with the classification of signal patterns in terms of generalized Hodgkin-Huxley models, the present work both classifies phenomena in terms of their generative mechanisms, and helps to define as structured data groups of phenomena that had heretofore seemed to be unrelated. We will develop the model in three stages and will indicate those data properties that can be explained at each stage. The first stage defines the basic gated pacemaker, whose properties already suffice to explain various circadian data. The second stage augments the gated pacemaker to include feedback generated by metabolic activity, which we think of as an index of fatigue. The analogous process in models of motivated behavior such as eating is a feedback signal due to satiety (Grossberg, 1982a, 1982b, 1982c). The third stage includes a slowly varying automatic gain control process. In models of motivated behavior, the analogous process describes how cues that are persistently associated with a motivated behavior become conditioned reinforcers that can thereupon modulate the activity cycle of that behavior. Our gain control process is not a "learning" process of the type that Pittendrigh and Daan (1976) criticized by saying (p. 248): "We see no utility in...treatment of circadian rhythms as being 'imprinted' on organisms...The structure of circadian pacemakers, including provision for some lability is completely encoded in DNA...". The gain control process is free from this criticism both because the basic gated pacemaker properties are independent of the gain control process and because all the processes in the model could be genetically specified.

The gated pacemaker model represents a single pacemaker system.

We have in mind the pacemakers that have been located in each of the two suprachiasmatic nuclei (SCN) of mammals (Moore, 1973, 1974; Moore-Ede, Sulzman, and Fuller, 1982). Various studies have indicated that these pacemakers drive the sleep, activity, feeding, and drinking cycles in mammals (Enright, 1980; Moore-Ede, Sulzman, and Fuller, 1982; Wever, 1979). In humans, the temperature cycle can be desynchronized from the sleep cycle, thereby suggesting that the temperature cycle is driven by a separate pacemaker system (Wever, 1979).

Recent models have focused on how these two distinct pacemaker systems are coupled (Enright, 1980; Kronauer, Czeisler, Pilato, Moore-Ede, and Weitzman, 1982; Wever, 1979). In these models, the individual pacemakers are chosen for convenience and simplicity, but do not admit a detailed physiological interpretation. For example, the Kronauer et al (1982) and Wever (1962, 1975) models consist of a pair of coupled van der Pol equations. The model of Kawato and Suzuki (1980) consists of a pair of coupled FitzHugh-Nagumo equations. The more abstract model of Daan and Berde (1978) describes a pacemaker entirely in terms of its period, phase, and phase shifts. Our analysis complements these contributions on the coupling of formal oscillators by explicating the dynamics of a single pacemaker. Those results about coupled oscillators which are insensitive to the detailed properties of the individual oscillators will carry over to the case where gated feedback networks are the oscillators to be coupled.

We, however, argue that some properties which have heretofore been assumed to necessarily follow from the coupling between oscillators can be explained by internal properties of a single oscillator, notably rhythm splitting and long-term after-effects. This claim does not deny the existence of coupling between distinct sleep and temperature systems. Nor does it deny the existence of distinct pacemakers in each of the two suprachiasmatic nuclei of a mammal. Rather we show

how properties which cannot be explained without coupling between classical oscillators can be explained by a single gated oscillating system, and that some properties which have not been explained by coupling between classical oscillators can be explained by a single gated oscillator. At the very least, these results show that further argument is needed to conclude that a coupling between oscillators generates a data property when the individual oscillators are poorly characterized.

III.3. <u>Some Circadian Phenomena and Gated Pacemaker Properties</u>

This section summarizes some of our model's explanations of circadian data. A more complete exposition is contained in our other articles (Carpenter and Grossberg, 1982a, 1982b). In our analysis, certain circadian properties are attributed to the gated pacemaker. Other phenomena are understood as due either to metabolic feedback or to slow gain control. Some important properties, such as slowly evolving split rhythms, require the entire system. Each level of analysis provides specific predictions about the anatomy and physiology of the circadian system. Several of these predictions challenge prevailing assumptions in circadian models.

A) <u>Competition Between On-Cells and Off-Cells</u>

The basic model consists of on-cell/off-cell pairs, or dipoles, which mutually inhibit one another. The on-cell drives observable activity, such as wheel-turning. Light is hypothesized to influence the endogenous circadian cycle by differentially exciting the on-cells or off-cells of the dipole, depending on whether the model depicts a diurnal or nocturnal animal. These hypotheses are consistent with the observations that electrically stimulating the optic nerve, or stimulating the retina by light, excites some cells of the suprachiasmatic

nuclei while inhibiting others (Groos and Mason, 1978; Groos and Hendriks, 1979; Lincoln, Church, and Mason, 1975; Moore-Ede, Sulzman, and
Fuller, 1982; Nishino, Koizumi, and Brooks, 1976). When these data are
interpreted in terms of a van der Pol or other formal oscillator, an
anatomical conclusion is drawn that differs from our own viewpoint in
the following way.

B) Phase Resetting in Diurnal and Nocturnal Animals

In the gated pacemaker model, the assumption that light inputs
excite the on-cells of diurnal animals and the off-cells of nocturnal
animals leads to the expected day-activity of diurnal animals and
night-activity of nocturnal animals. This assumption also implies that
light resets the phase of both diurnal and nocturnal gated pacemaker
models in a similar way, as demanded by the data (Pittendrigh, 1960,
1974).

For example, during the "early subjective night" of a model diurnal animal, a light pulse exciting the on-cell prolongs its active
phase, delays the rest cycle, and thereby creates a phase delay. During the "early subjective night" of a model nocturnal animal, a light
pulse exciting the off-cell prolongs its active phase, delays the en-
suing activity cycle, and again creates a phase delay. During the
"late subjective night" of a diurnal animal, a light pulse exciting
the on-cell induces a premature onset of on-cell activity, thereby
causing a phase advance in the onset of activity. During the "late
subjective night" of a nocturnal model animal, a light pulse exciting
the off-cell induces a premature onset of off-cell activity, thereby
causing a phase advance in the onset of rest which, in turn, causes a
phase advance in the onset of activity. For both diurnal and nocturnal model animals, a light pulse during the "subjective day" has little effect. Thus the characteristic phase response curves of both di-

urnal and nocturnal model animals are similar, as also occurs _in vivo_.
Thus if a gated pacemaker exists in the SCN, then the phase response
curves of both diurnal and nocturnal animals can be explained by SCN
dynamics.

By contrast, if a van der Pol oscillator is used to model the
pacemaker, then it is difficult to see how the diurnal/nocturnal distinction could be built in until after the SCN level, as Moore-Ede et
al (1982, p.81-82) realized: "The circadian systems of diurnal and
nocturnal species must be organized differently to account for the
dramatic differences in the phase relationships of their rhythms to
the light-dark cycle [i.e., day-active vs. night-active]. It is possible that the differences lie in the coupling between zeitgeber and
pacemaker. However...the similarities between nocturnal and diurnal
species in the way that light resets circadian pacemakers [i.e., the
phase response curves] make it more likely that the difference in the
phase relationships of the rhythms of nocturnal and diurnal animals
actually depends on differences in the coupling mechanisms between the
circadian pacemaker and the rhythms it drives."

If the transmitter gates which our model hypothesizes to exist in
the SCN could be parametrically excited or inhibited during phase resetting experiments, then predictable changes in the phase response
curves would be generated that could not be explained by a formal oscillator model.

C) Suppression of Circadian Rhythm by Steady Bright Light

Another property of a gated pacemaker that is often not discussed
in formal models of coupled oscillators concerns the parametric response to increases in a steady background light level. Aschoff (1979,
p.238) writes: "At high intensities of illumination circadian systems
often seem to break down, as primarily exemplified by arhythmicity in

records of locomotor activity. Data supporting this statement have not often been published but the phenomenon is familiar to everyone working in the field."

In a gated pacemaker, sufficiently high steady light quenches the circadian rhythm. Activity is then determined by competition between motivated behaviors that are modulated by aperiodic environmental cues.

D) Period Doubling and Biorhythms

When humans live in caves for long periods in dim steady light, their circadian rhythm occasionally drifts towards a forty-eight hour day (Jouvet, Mouret, Chouvet, and Siffre, 1974). A period doubling phenomenon can also occur in the basic gated pacemaker model without light input. A normal period is achieved using the same choice of parameters in response to periodic light. Slow modulations of activity on a time scale much longer than a day can also occur. These phenomena are described in greater detail in Section III.6.

E) Split Rhythms and Metabolic Feedback

Pittendrigh (1960) first noted, and recognized the importance of, the phenomenon of split rhythms, whereby a nocturnal animal with a single daily activity cycle in the dark may generate an activity cycle which splits into two components in constant light. In recent years, numerous examples of split rhythms have been discovered. Hoffman (1971) described a diurnal animal (Tupaia belangeri) whose rhythm splits when the level of illumination is reduced; and Gwinner (1974) noted that the hormone testosterone induces split rhythms in starlings. Pittendrigh (1974) also noted that "Many animals tend to be bimodal in their activity pattern"(p.450) even when the activity pattern does not split.

Since Pittendrigh's original observations, many circadian models have adopted Pittendrigh's assumption that split rhythms are due to a pacemaker consisting of two or more coupled oscillators which drift out of phase when the split occurs:

"The circadian pacemaker for the activity cycle [α] comprises two separable oscillators, one responsible for the N [night, or earliest] component of α and the other for its M [morning, or later] component." (Pittendrigh, 1974, p.450).

"The findings reported here strongly suggest that the rhythms of locomotor activity in Tupaias is controlled by two coupled oscillators (or two groups of oscillators), which may have two stable phase relationships." (Hoffmann, 1971, p.142).

"Several recent investigations in mammals have made it virtually certain that the daily rhythm of gross locomotor activity...is governed by at least two coupled oscillators. This is strongly suggested by the observation that under certain conditions of constant illumination the rhythm of locomotor activity may 'split' into two distinct components..."(Gwinner, 1974, p.72).

"...Most of the behaviour of the rhythms observed may also be interpreted as complex responses of a single basic oscillator...The most compelling evidence for a two-oscillator system in vertebrates is the occurrence of 'splitting' of free-running activity rhythms into two distinct components." (Daan and Berde, 1978, p.298).

"...only a multi-oscillator arrangement could account for all the various behaviors of the mammalian timing system...Splitting of the activity pattern of rodents (Pittendrigh, 1960) was one of the earliest indications of multiple, potentially independent oscillators in mammals." (Moore-Ede, Sulzman, and Fuller, 1982, p.117-118).

We challenge the assumption that split rhythms necessarily imply the existence of in-phase/out-of-phase oscillators by explicitly demonstrating the existence of both split rhythms and bimodal activity patterns in a different type of model. At best, the traditional two-oscillator model now requires further proof.

In the gating model, split rhythms are caused by metabolic feedback due to activity. In this context, bimodal activity patterns are generated as follows: The pacemaker excites the on-cell and thereby initiates activity, such as wheel-turning. Activity causes a build-up of metabolic feedback to the off-cell. This feedback enables the off-cell to partially inhibit the on-cell earlier than the on-cell transmitter gate's relatively large value would otherwise allow. Wheel-turning thereupon slows or ceases. The metabolic feedback has an opportunity to dissipate during this rest period. The on-cell is thereby disinhibited and its potential is revived by the still relatively large value of the on-cell transmitter gate. Activity thereupon increases until the combination of increased metabolic feedback and off-cell activation by the pacemaker bring on rest and sleep. For the same model animal, an environment that depresses on-cell potential, such as constant light in a nocturnal animal, turns the rest period into the full-fledged sleep period of the split rhytym.

Our explanation thus suggests that a nonspecific effect on the pacemaker, say due to substances in the bloodstream, can cause split rhythms. This explanation is consistent with electron microscopic evidence that some SCN cells are clustered in direct apposition to the walls of blood capillaries (Moore, Card, and Riley, 1980; Card, Riley, and Moore, 1980). As Moore-Ede et al (1982) note, these cells "may act as receptors, sensing hormonal signals from elsewhere" (p.172). Indeed, it has been observed that injection of hormones can induce split rhythms (Gwinner, 1974). Our hypothesis is also supported by the fact that split rhythms are observed in higher organisms, although, in theory, it is possible for depletion of a chemical in a lower or unicellular organism to have a similar effect.

The metabolic feedback term in our model explicates the intuitive idea that activity per se, or other metabolically mediated processes, can influence our need to rest. The model also indicates how subtle the interaction between the underlying pacemaker and contingent activity can be. Because the metabolic feedback process rises and falls

in the model on an "ultradian" time scale that is significantly short-
er than the 24-hour day, it helps to account for the approximately
6-, 8-, or 12-hour components of activity sometimes seen, for example,
when a nocturnal animal is put in the dark after its circadian rhyth-
micity has been eliminated by leaving it in steady light (Pittendrigh,
1960, p.172).

F) Unilateral Lesions of the SCN Abolish Split Rhythms: The
   Internal Zeitgeber

When a nocturnal animal such as a golden hamster is maintained in
constant light, a split rhythm can develop. Surgical ablation of one
of the two SCN in the hamster eliminates the split rhythm (Pickard and
Turek, 1982). The ablation also causes a total reduction in activity
and a greater temporal diffuseness of activity.

These data imply that a functional relationship exists between
the two SCN that helps to synchronize as well as to split the hamster's activity rhythm. The nature of this functional relationship requires close scrutiny. Pickard and Turek (1982) wrote that more than
one interpretation is possible: "...the two SCN oscillators...might
normally be coupled, but this coupling might be altered under...the
split condition...Another possibility is that a set of interacting
pacemakers may reside in each SCN, and the loss of the split rhythm
may be a consequence of the total number of these oscillators destroy-
ed; whether or not the destruction is unilateral may not be important"
(p.1121). The former interpretation of two oscillators going out of
phase due to a change in their mutual coupling is the more familiar
one. To explain how removal of one SCN creates diffuse overall activity, this interpretation would need to suppose that several oscillators
exist in each SCN and that, although the oscillators within either SCN
do not mutually interact, the oscillators between the SCN do interact,

probably via an inhibitory coupling.

The gating pacemaker model explains both the abolition of the split as well as the reduction and diffuseness of subsequent activity without supposing that individual oscillators are differentially coupled within and between the two SCN. The model does this by explicating how a change in total activity causes the observed effects by reducing the metabolic feedback received by each cell in the remaining SCN. This explanation proceeds as follows.

Before an SCN is removed, every on-cell in each SCN contributes to a total excitatory signal that supports the observed activity level. (See function $G(x_1(t))$ in equation (65) of Section III.7). This total excitatory signal determines the total amount of metabolic feedback. Every off-cell in each SCN receives this _total_ metabolic feedback signal. That is, the metabolic feedback signal is distributed _nonspecifically_ to all the off-cells. In Section (E) above, we explained how this metabolic feedback signal can cause a split rhythm.

When one SCN is extirpated, the total number of on-cells that can generate activity is cut in half. Consequently, the total metabolic feedback signal to _each_ surviving off-cell is soon also significantly reduced. This fact immediately indicates how the split rhythm is abolished, since in our model the split rhythm is ascribed to a relatively large metabolic feedback signal.

The diffuseness of activity is then explained in our model as it would be in a coupled SCN model. We assume that the remaining SCN contains several gated pacemakers which, in the absence of entraining signals, eventually get desynchronized. By contrast with a coupled SCN model, we suggest that the entraining signal may be received via the bloodstream; hence, its nonspecific character. To emphasize the fact that this metabolic feedback signal, which acts like a forcing function analogous to light, can be controlled by internal factors other than the pacemakers themselves, we call it an __internal zeitgeber__.

G) Split Rhythm After-effects: Slow Gain Changes

When a nocturnal animal, such as a golden hamster, is placed in steady light, it may take a month or more for a split rhythm to evolve (Pittendrigh, 1974, p.449). In the present model, the slowness of the split onset is due to the action of a slow gain change that is analogous to the change in a cue's conditioned reinforcing efficacy in models of motivated behavior (Grossberg, 1982a, 1982b, 1982c). More precisely, the slow gain change is a process with two properties: its strength increases slowly as a function of on-cell activity; it acts as an excitatory signal to the on-cell that is proportional to its strength when the model animal is active. Otherwise expressed, the slowly varying gain process acts to gate an excitatory signal to the on-cell. This gating action is functionally distinct from the gating action that generates the underlying pacemaker oscillation.

Speaking intuituvely, the slowly varying gain signal causes the slow onset of the split as follows. A split rhythm is caused when metabolic feedback can get sufficiently large relative to the on-cell potential during an activity cycle. At the time when the model animal is placed in steady light, the slow gain signal is relatively large. It thereby enhances the on-cell potential during an activity cycle. After the model animal is placed in steady light, the light acts to excite the off-cell and thus tends to inhibit the on-cell. At first, the relatively large gain signal partially offsets this reduction in on-cell activity. Gradually, however, the gain signal senses the average reduction in the on-cell potential. The gain signal slowly decreases as a result, and the on-cell activity decreases further, on the average. As this progressive decrease in average on-cell activity proceeds, the metabolic signal gradually causes the rhythm to split.

When an animal whose rhythm is split is placed back into its original steady light environment, its rhythm can remain split indefinitely (Hoffmann, 1971). In our model, the split rhythm occurs when
the gain signal is sufficiently small to allow metabolic feedback to
split the on-cell activity cycle. Under certain circumstances, the
average on-cell activity thereby decreases. When this occurs, the gain
signal also tends to remain small. The splitting mechanism hereby
tends to perpetuate itself unless a sufficiently potent and sustained
counteracting influence is imposed, as Hoffmann (1971) also found.

Many authors interpret the slow onset and offset of split rhythms
as a desynchronization phenomenon. In our theory, the slow onset and
offset are attributed to the same mechanism that we use to explain
after-effects, as in Section (I) below. In fact, our theory explains a
variety of slowly varying processes using just this gain control mechanism. In particular, this mechanism has been used to numerically simulate several of the different types of split rhythms that are found
in the data (Pittendrigh, 1974, p.449).

H) Aschoff's Rule and Its Exceptions: Paradoxical Results on
   After-effects

Additional support for the existence of a slow gain control process that reflects average activity comes from a correlative analysis
of three types of data: Aschoff's rule and its exceptions, split rhythms in diurnal and nocturnal animals, and long-term after-effects on
period subsequent to phase leads and lags caused by light pulses. In
the experimental literature, no correlation has been drawn between
these three types of phenomena. Our model rationalizes the split rhythm data using the mediating hypothesis that a slow decrease in the
gain control signal tends to make split rhythms more likely, whereas
a slow increase in the gain control signal tends to make split rhythms

less likely.

Aschoff's rule notes the tendency in many nocturnal mammals for an increase in steady light to cause an increase in period and a decrease in activity, and in some diurnal mammals for an increase in steady light to cause a decrease in period and an increase in activity (Aschoff, 1979). Exceptions to this rule frequently occur. In a different set of experiments, some diurnal mammals show split rhythms caused by an increase in light; others have rhythms which split in response to a decrease in light. Also, in some diurnal mammals, a transient lengthening of period during a phase shift is followed by a free-running rhythm of shortened period, whereas after a transient shortening of period, a lengthened free-running period is observed (Kramm, 1971). We correlate and unify such apparently unconnected observations using the properties of our gain control mechanism in another article of this series (Carpenter and Grossberg, 1982a).

I) Frequency After-effects

Pittendrigh (1974) wrote: "In our laboratory, we have found after-effects on the frequency of freerunning [in the dark] rhythms following: (1) phase shifts induced by light signals; (2) entrainment by cycles whose period is near the limit of entrainment...; (3) exposure to constant light; (4) change in photoperiod" (p. 441). Pittendrigh also noted that no mathematical model could yet explain these widespread phenomena. We have successfully simulated all of these phenomena using the slow gain control mechanism.

We will indicate below how the gain control mechanism generates one of these types of after-effects. In general, all of the cited manipulations which cause after-effects alter the total activity level, and thus the size of the gain signal in the model. The gain signal, and therefore the circadian period, may gradually return to its prior

level or may maintain a new level indefinitely, depending on the experiment.

An experiment in which photoperiod was manipulated will now be
summarized (Pittendrigh, 1974, p.438). The photoperiod is the total
duration of light within one circadian cycle. The experiment was done
on a nocturnal rodent, the deermouse. First light was turned on periodically for one hour, followed by twenty-three hours of dark, for
sixty days. Then the animal free-ran in the dark for thirty days.
Next the animal was periodically exposed to eighteen hours of light
followed by six hours of darkness for fifty days, after which the animal again free-ran in the dark until the end of the experiment. The
total periods and activity levels during the two free-run intervals
were different and maintained themselves for at least thirty days. The
activity level after one-hour light pulses exceeded that after
eighteen-hour light pulses. The period after one-hour light pulses
also exceeded the period after eighteen-hour light pulses. Both of
these effects were found in our numerical simulation despite the fact
that the animal does not obey Aschoff's rule in this experiment.

The reason for the difference in activity levels is easily explained in terms of slow gain change. The difference in periods is
due to a complex interaction that is not easily described in words.
To see why free-running activity level decreases as light duration
during the preceding photoperiods increases, note that an increase of
light duration in a nocturnal animal causes a decrease in average on-
cell activity, which in turn causes a gradual decrease in the gain
signal. During the subsequent free-run interval, the smaller gain signal supports a smaller average level of on-cell activity. In all, an
increase in light duration during the photoperiod causes a sustained
decrease in activity during the subsequent free-run period.

III.4. The Gated Pacemaker

The gated pacemaker is described by a four-dimensional dynamical system, just as in the case of the Hodgkin-Huxley (1952) model. Whereas the Hodgkin-Huxley model contains one potential to which three auxiliary variables are coupled, the gated pacemaker model possesses two mutually interacting potentials to each of which is coupled a slow gating process. More precisely, the model describes interactions between two pairs of variables, (x_1, z_1) and (x_2, z_2). Each x_i is a "fast" variable that represents the voltage (or activity) of a cell (or cell population) v_i, $i=1,2$. Each z_i is a "slow" variable that represents the amount of stored transmitter in an excitatory feedback pathway from v_i to itself whose signals are gated by z_i, $i=1,2$. In particular, the z_i's correspond to the z variable in the gating model of Section II, equation (28).

The cell(s) v_1 is an on-cell and the cell(s) v_2 is an off-cell. Such on-cell/off-cell pairs, or dipoles, are widely found in the nervous system (Thompson, 1967). The on-cell and off-cell characteristics of v_1 and v_2 are due to the following constraints.

The potentials $x_1(t)$ and $x_2(t)$ mutually inhibit one another. The input J(t) which represents the (transduced) intensity of light excites the on-cell when the model represents a diurnal animal, but excites the off-cell when the model represents a nocturnal animal (Figure 18). In both diurnal and nocturnal model animals, the on-cell output energizes behavioral activity. Both the on-cell and the off-cell also receive an equal tonic input I that represents the arousal level of the dipole. Our model of the pacemaker in a suprachiasmatic nucleus consists of a family of these on-cell off-cell dipoles. Many of our results can be derived from the properties of a single such dipole.

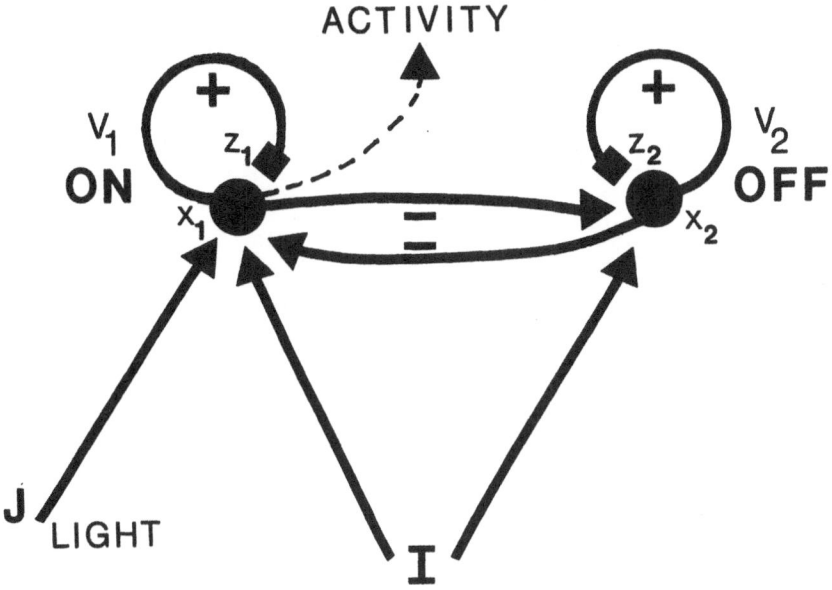

Figure 18: On-cell/off-cell anatomy with positive gated feedback and mutual inhibition. I is the tonic arousal input. J(t) represents the light input, which enters the on-cell in the diurnal model and which enters the off-cell in the nocturnal model.

The equations which describe the dynamics of the anatomy in Figure 18 are exactly analogous to the membrane equation (34) and the gating equation (28) from Section II:

$$C_0 \frac{dV}{dt} = (V^p - V) g^p + (V^+ - V) g^+ + (V^- - V) g^- \qquad (34)$$

and

$$\frac{dz}{dt} = A(B-z) - S(t) z. \qquad (28)$$

The on-cell and off-cell gating equations are thus

$$\frac{dz_1}{dt} = A(B-z_1) - f(x_1)z_1 \tag{52}$$

and

$$\frac{dz_2}{dt} = A(B-z_2) - f(\dot{x}_2)z_2 \tag{53}$$

where $f(x_i)$ is proportional to the excitatory feedback signal transmitted by cell v_i, i=1,2. Our analysis considers signal functions of
the form

    i) <u>Linear Above Threshold</u>

$$f(w) = \begin{cases} w \text{ if } w > 0 \\ 0 \text{ if } w \leq 0 \end{cases} \tag{54}$$

or

    ii) <u>Sigmoid</u>

$$f(w) = \begin{cases} \dfrac{w^2}{\alpha^2+w^2} & \text{if } w > 0 \\ 0 & \text{if } w \leq 0 \end{cases} . \tag{55}$$

The membrane equations for diurnal x_1 and x_2 are:

$$C_0 \frac{dx_1}{dt} = (x^p-x_1)g^p + (x^+-x_1)[I+J(t)+Cf(x_1)z_1] - (x^--x_1)Dg(x_2) \tag{56}$$

and

$$C_0 \frac{dx_2}{dt} = (x^p-x_2)g^p + (x^+-x_2)[I+Cf(x_2)z_2] - (x^--x_2)Dg(x_1) . \tag{57}$$

In (56) and (57), the constants x^p, x^+, and x^- are the passive, excitatory, and inhibitory saturation points, respectively. The excitatory
conductance of x_1 is the sum of a tonic arousal input I, a light input
J(t), and a positive gated feedback signal $Cf(x_1)z_1$. By contrast, the

excitatory conductance of x_2 does not contain a light input; in a model of a nocturnal animal, $J(t)$ is added to the excitatory conductance of x_2. The inhibitory conductance of x_1 is proportional to a feedback signal $g(x_2)$ from v_2. Typically $g(w)$ is chosen as in (54) or (55). The inhibitory conductance of x_2 is proportional to a feedback signal $g(x_1)$ from v_1. The two inhibitory conductances, taken together, express the mutual inhibition of v_1 and v_2.

Without loss of generality, we set $x^p = 0$ and put equations (52)-(57) in dimensionless form as follows:

$$\frac{dx_1}{dt} = -x_1 + (1-x_1)[C_1+J(t)+C_2f(x_1)z_1] - (x_1+C_3)C_4g(x_2) , \qquad (58)$$

$$\frac{dx_2}{dt} = -x_2 + (1-x_2)[C_1+C_2f(x_2)z_2] - (x_2+C_3)C_4g(x_1) , \qquad (59)$$

$$\frac{dz_1}{dt} = C_5(1 - z_1 - C_6f(x_1)z_1) , \qquad (60)$$

and

$$\frac{dz_2}{dt} = C_5(1 - z_2 - C_6f(x_2)z_2) , \qquad (61)$$

where $C_1,\ldots,C_6$ are positive dimensionless constants; $x_1(t)$ and $x_2(t)$ are dimensionless variables which remain between $-C_3$ and 1; and $z_1(t)$ and $z_2(t)$ are dimensionless variables which remain between 0 and 1. Note that if the light input $J(t)$ is identically zero, then the pairs of equations for (x_1,z_1) and (x_2,z_2) are symmetric.

Further commentary is needed to characterize the light input $J(t)$. We assume that when light is on and the model animal is awake, then $J(t)$ equals the experimentally controlled light intensity. If the model animal is asleep, then $J(t)$ equals a constant fraction of the experimentally controlled light intensity. The model animal goes to sleep when the on-cell potential $x_1(t)$ is smaller than a prescribed constant. In terms of equations (65) and (68) below, we assume that the animal goes to sleep when $G(x_1(t)) \le \theta C_9$ where $0 < \theta < 1$.

III.5. <u>Genesis of Unforced Pacemaker Oscillations: Strength of</u> <u>Inhibitory Coupling</u>

This section and the next consider some of the types of oscillations that can occur within an unforced gated pacemaker in the dark. In this situation, the light input $J \equiv 0$ in (58). Consequently equations (58) and (59) for x_1 and x_2 and equations (60) and (61) for z_1 and z_2 are symmetric, and could represent pacemaker activity of either a nocturnal or a diurnal animal. This section describes how oscillations depend on the strength of the inhibitory coupling constant C_4 in (58) and (59). The next section describes how oscillations depend on the choice of the threshold constant C_7 of the excitatory feedback function

$$f(w) = \begin{cases} \dfrac{w^2}{C_7^2 + w^2} & \text{if } w > 0 \\ 0 & \text{if } w \le 0 \end{cases} \tag{62}$$

when

$$g(w) = \begin{cases} w & \text{if } w > 0 \\ 0 & \text{if } w \le 0 \end{cases}. \tag{63}$$

In both cases, the choices (62) and (63) of signal functions are made. Surprisingly, parametric changes in C_7 cause totally different oscillatory waveforms than do parametric changes in C_4.

Choosing $C_4 = 0$ decouples the two potentials x_1 and x_2. In all of our numerical studies, this choice forces all the variables to approach limits $x_1(\infty)$, $x_2(\infty)$, $z_1(\infty)$, and $z_2(\infty)$ such that

$$x_1(\infty) = x_2(\infty) \text{ and } z_1(\infty) = z_2(\infty). \tag{64}$$

Such limits are said to occur on the diagonal. If $f(w)$ is chosen to be linear above threshold, as in (54), then it can be proved that this outcome is necessary. We have, moreover, not found any numerical examples that violate this outcome. Consequently, we can assert that decoupling the two potentials quenches unforced oscillations. This conclusion strongly distinguishes our pacemaker model from other models in the literature.

Figures 19-24 describe how the behavior of the unforced gated pacemaker changes due to parametric increases in the inhibitory coupling strength C_4. This parametric series shows just the behavior one might expect: for weak coupling (Figure 19), the limit is on the diagonal; as C_4 increases, a small amplitude periodic solution bifurcates from the critical point (Figure 20); further increases in C_4 cause the amplitude and the period of the periodic solution to increase (Figures 21 and 22); a still stronger coupling elicits a large amplitude relaxation oscillation (Figure 23); finally, a very strong coupling enables the cell (population) v_1 or v_2, whichever has the larger initial values, to win the competition (Figure 24). The oscillation is thereby quenched and a limit is approached off the diagonal. Thus in all nontrivial cases, the solution pairs $(x_1(t), z_1(t))$ and $(x_2(t), z_2(t))$ eventually become 180° out of phase.

In Figures 19-24, two types of graphical representation are used. The first type of representation depicts the 4-dimensional phase portrait by projecting both pairs (x_1, z_1) and (x_2, z_2) onto an (x,z) coordinate plane. Each state of the system is then represented by a pair of 2-dimensional points. The second representation plots the functions $x_1(t)$ and $z_1(t)$ for $t \geq 0$. Because the pairs (x_1, z_1) and (x_2, z_2) are eventually 180° out of phase, the plots of $x_2(t)$ and $z_2(t)$ can also be

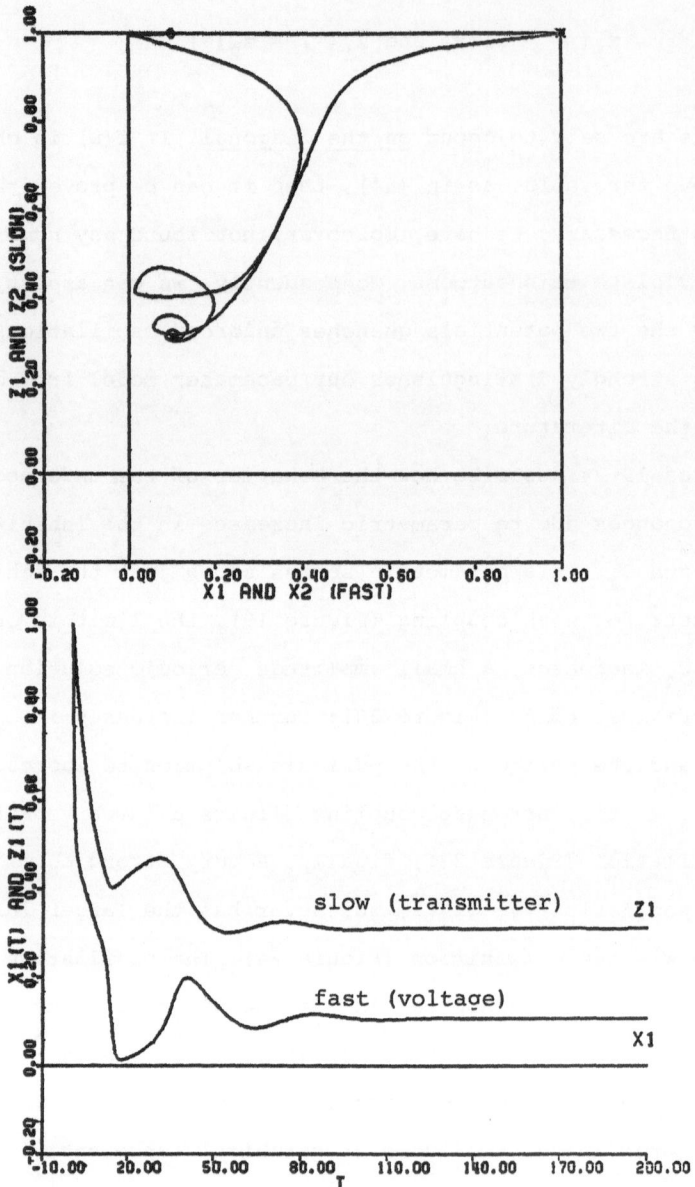

Figure 19: $C_4 = 2$. With weak inhibitory coupling, solutions go to a
limit on the diagonal. In Figures 19-24, $C_7 = .2$ and C_4
varies.

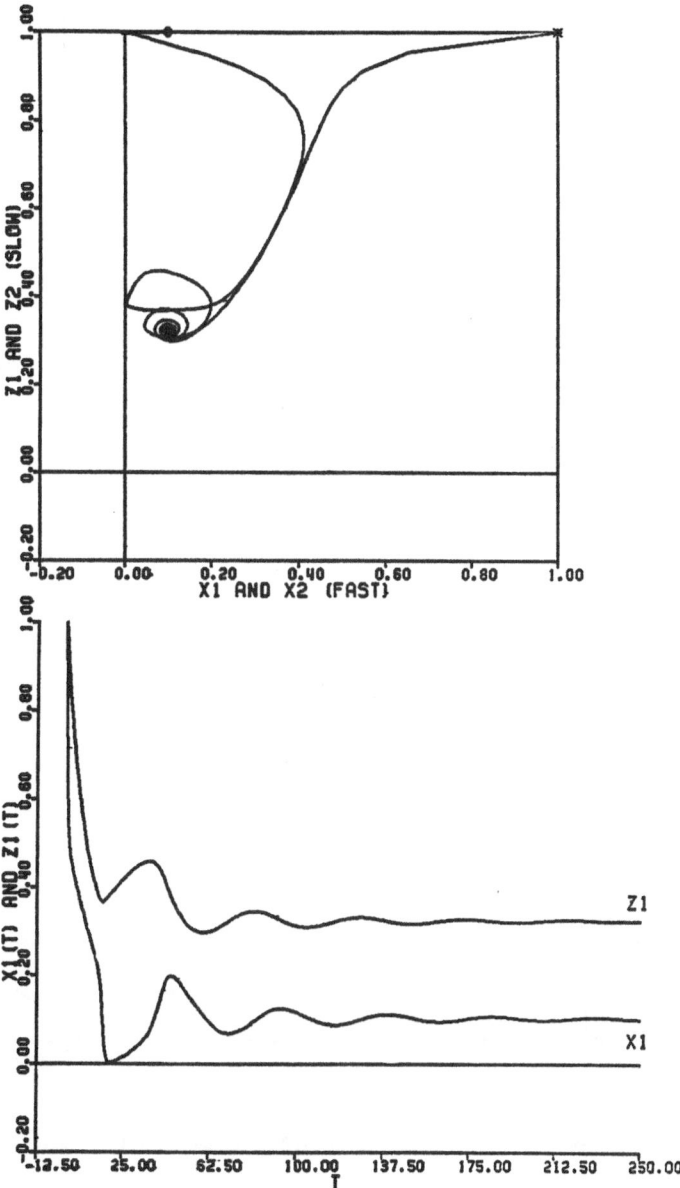

Figure 20: C_4 = 2.3. As coupling strength increases, a small amplitude periodic solution bifurcates from the critical point
on the diagonal.

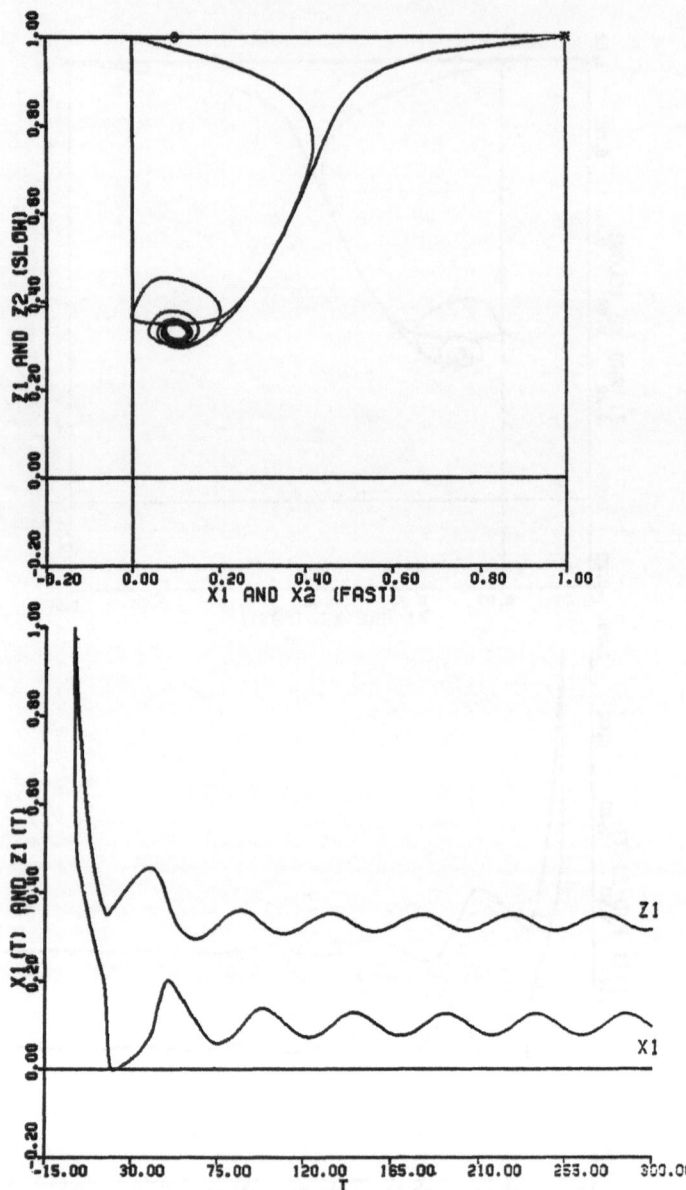

Figure 21: $C_4 = 2.5$. Amplitude and period increase.

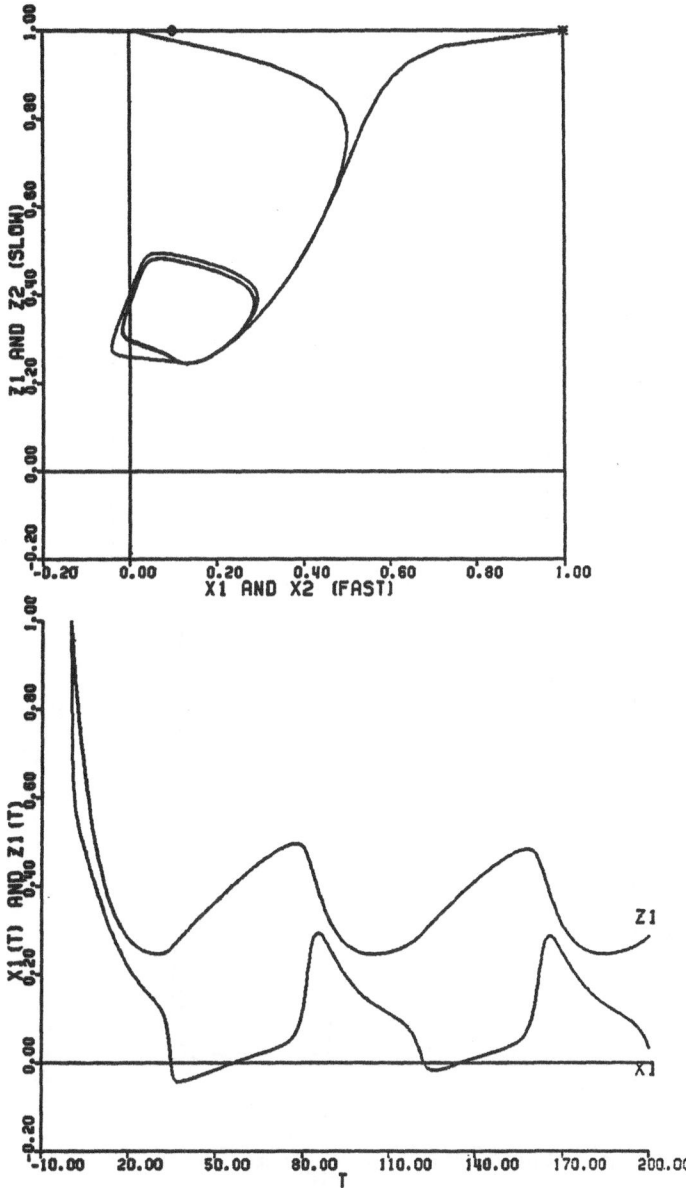

Figure 22: $C_4 = 5$. A large amplitude regular periodic solution.

178

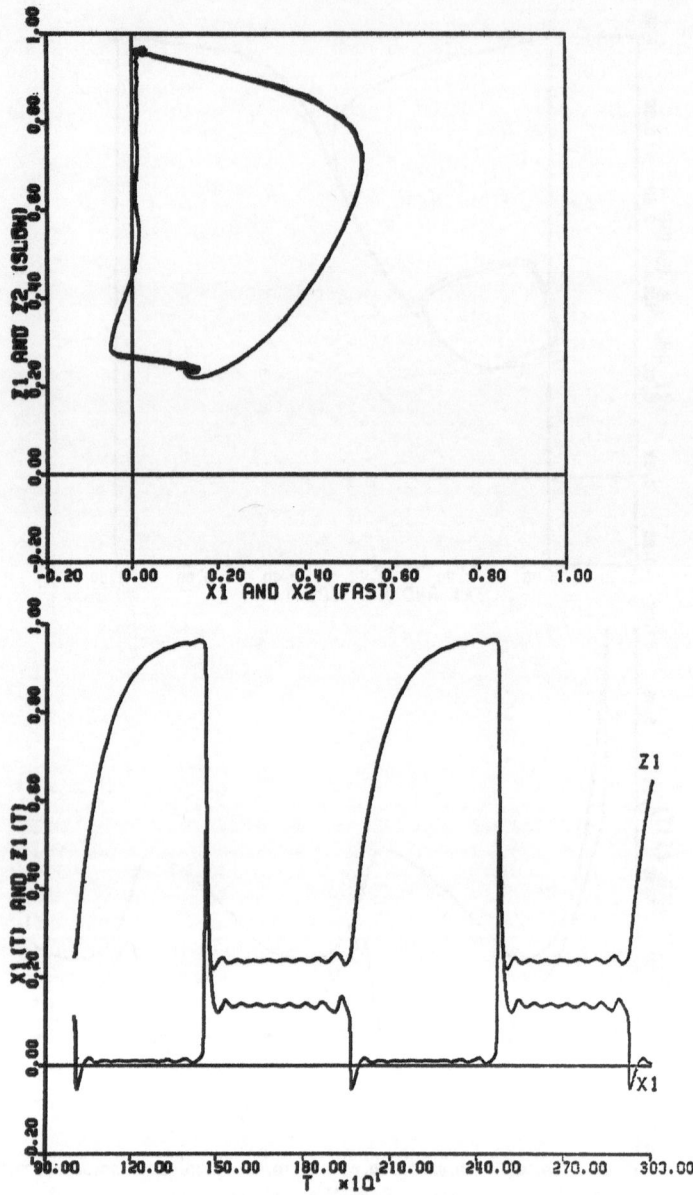

Figure 23: $C_4 = 6$. With strong inhibitory coupling, solutions oscillate near a plateau until there is a sudden switch, as if
a tug-of-war is taking place.

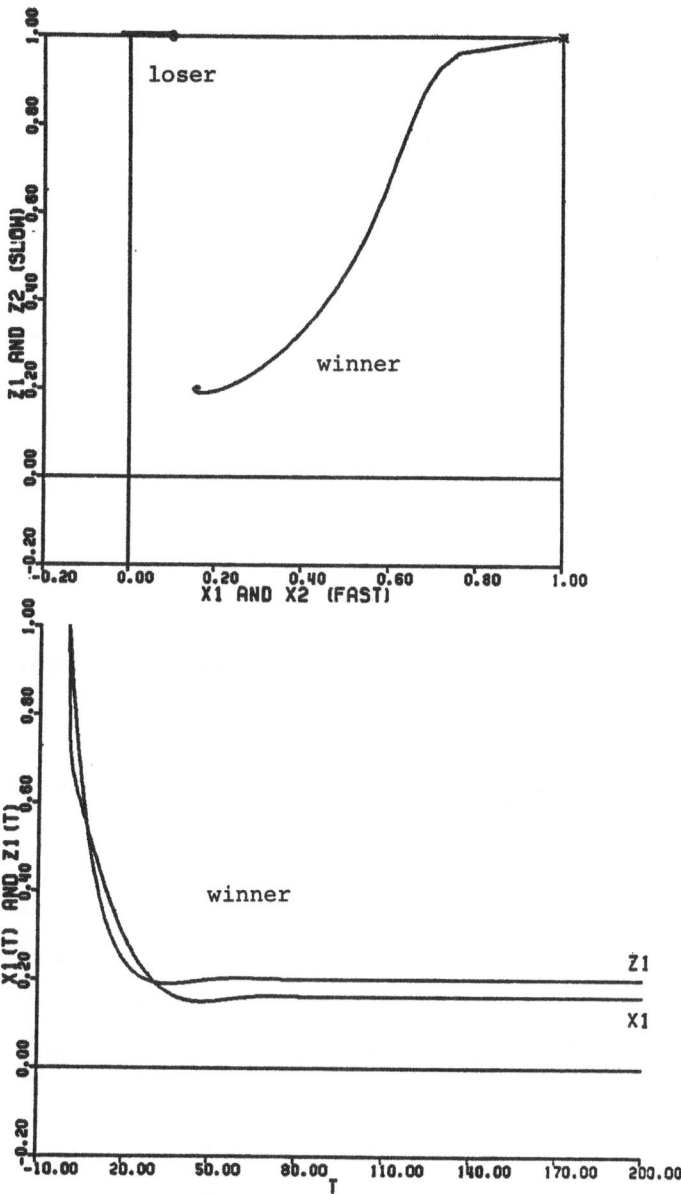

Figure 24: $C_4 = 7$. When inhibitory coupling is sufficiently strong, the pair (x_1, z_1) or (x_2, z_2) with the initial advantage wins the tug-of-war, and solutions go to a limit off the diagonal.

inferred.

The illustrated dynamics are robust over a wide range of parameters. Physical considerations have guided some constant parameter choices. For example, we let C_3 =.1 in (58) and (59) because C_3 is the ratio of $|V^-|$ to V^+. In vivo, V^+ often stands for the saturation point of a Na^+ channel and V^- stands for the saturation point of a K^+ channel such that $V^+ >> V^-$. In the Hodgkin-Huxley (1952) model, for example, $C_3 \cong .1$. Also the "slow" rate C_5 of gate accumulation in (60) and (61) is chosen small relative to the "fast" unit rate of potential decay in (58) and (59). In particular, we choose C_5 = .01.

III.6. Period Doubling, Slowly Modulated Irregular Periodic Waves, and Chaos

Parametric changes in the signal threshold C_7 in (62) cause a dramatically different and novel sequence of oscillatory patterns. We start with the system illustrated in Figure 22. Here the choices C_4 = 5 and C_7 = .2 caused large-amplitude oscillations.

If C_7 is chosen very small, the signal function $f(w)$ in (62) makes a sharp jump from 0 to 1 as w increases. If C_7 is large, this signal function increases gradually from 0 to 1 as w increases. Given any choice of $C_7 > 0$, $f(C_7) = \frac{1}{2}$. In the parametric series in Figures 19-24, C_7 = .2, which is a moderate level because x_1 and x_2 can oscillate between -.1 and 1.

Figure 25 depicts the same system as Figure 22, but at large times. Consequently, the projected phase portrait looks like a single periodic solution, although really the two pairs (x_1, z_1) and (x_2, z_2) traverse this closed curve 180° out of phase. As C_7 increases from .2 to .35, an interval of C_7 values is reached wherein period doubling occurs (Figure 26). In other words, as C_7 is increased from .2 to .35, the regular periodic solution of Figure 25 is gradually deformed in

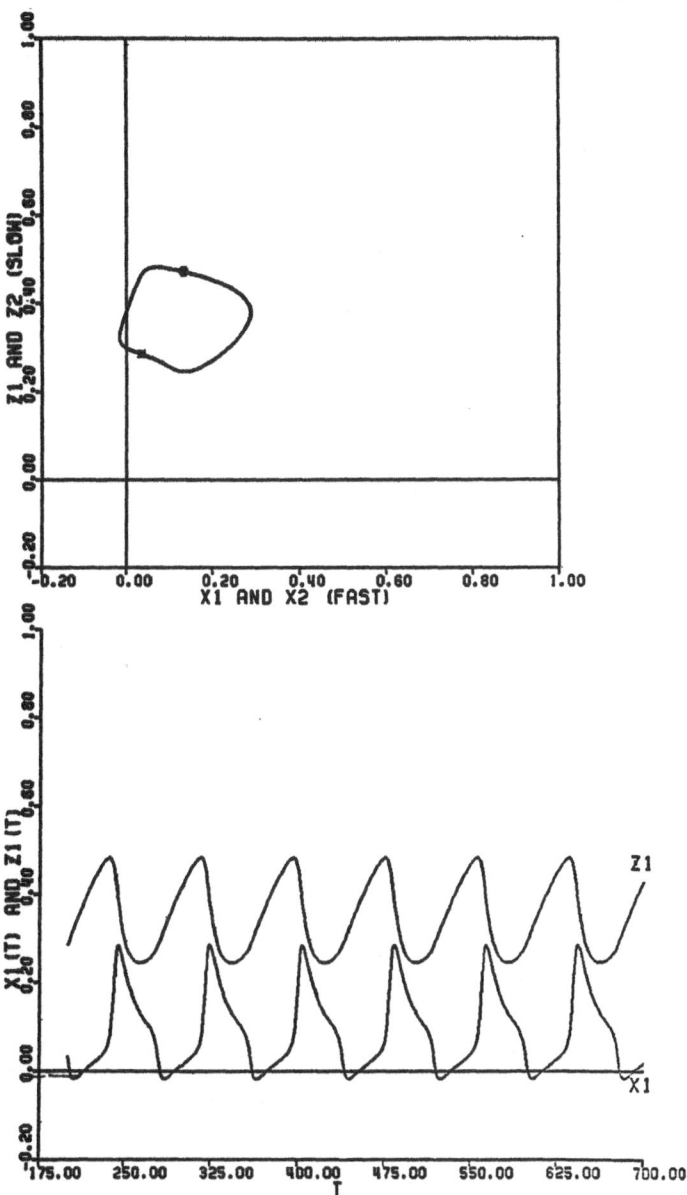

Figure 25: C_7 = .2. The large amplitude periodic solution of Figure 22, at later times. In Figures 25-29, C_4 = 5 and C_7 varies.

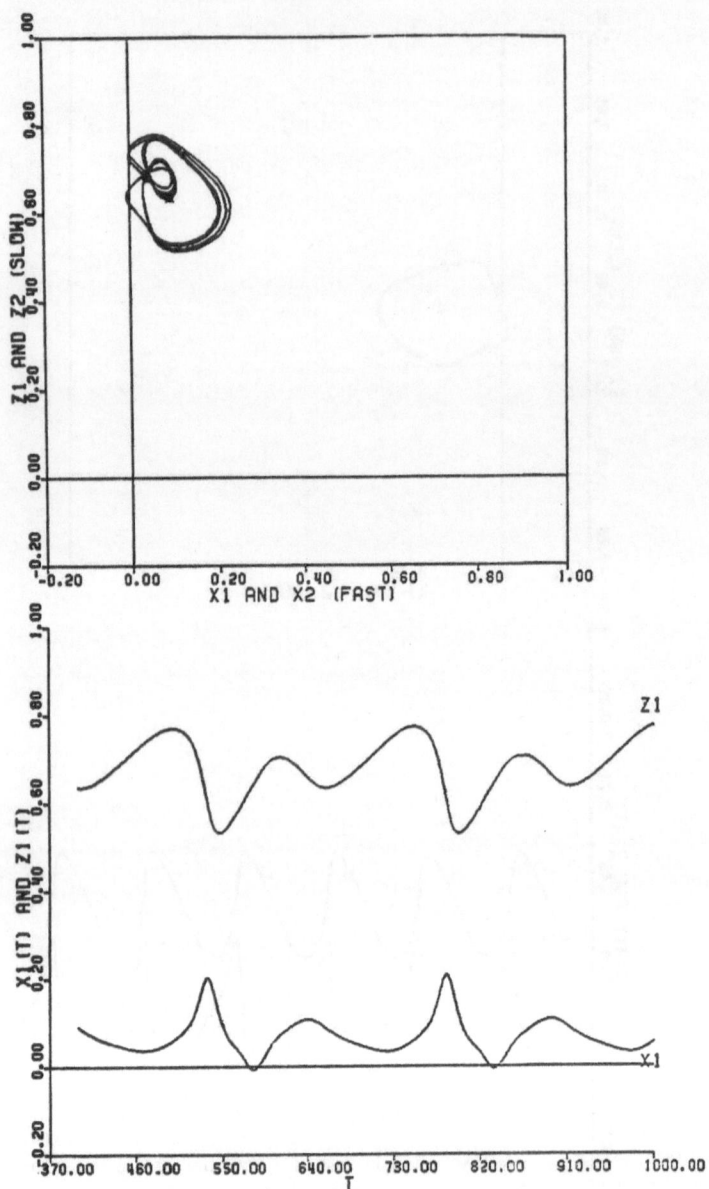

Figure 26: $C_7 = .35$. A period-doubled solution.

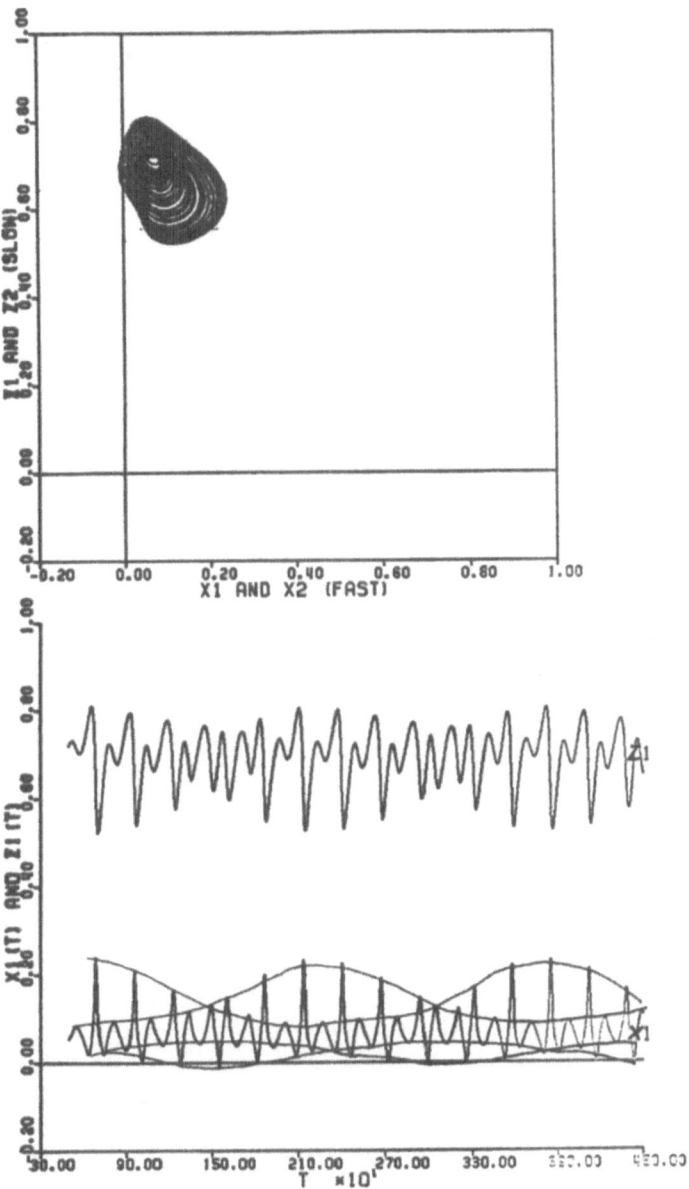

Figure 27: C_7 = .355. A slowly modulated period-doubled solution.
The modulation is itself periodic.

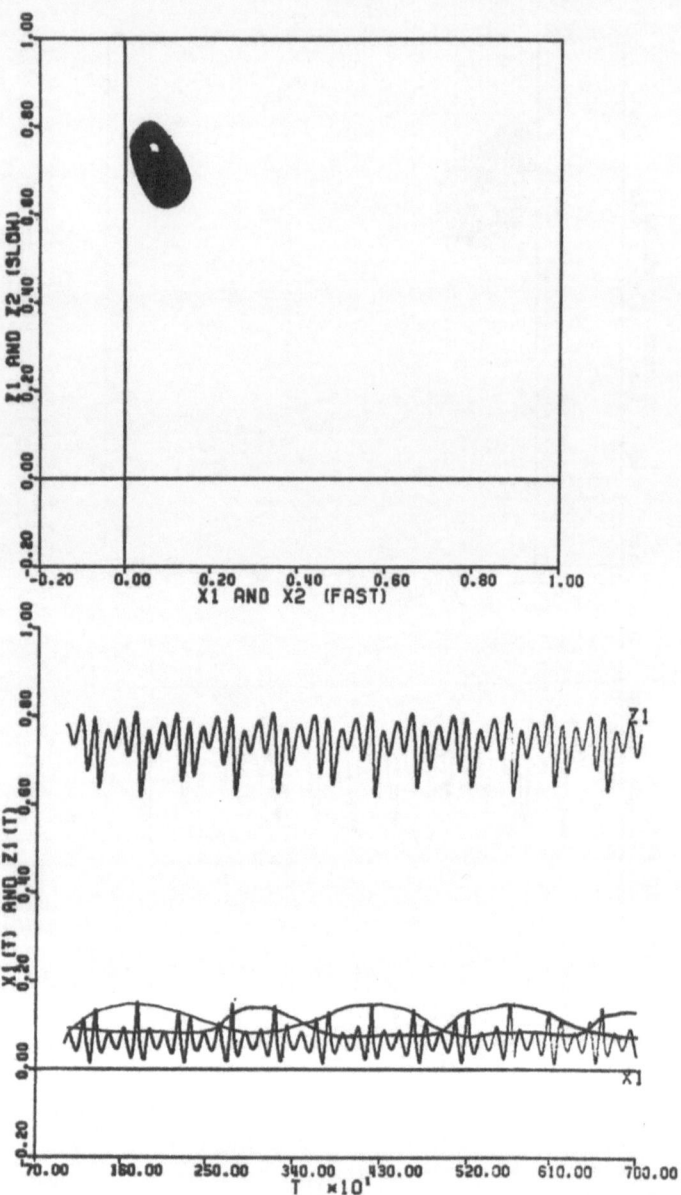

Figure 28: C_7 = .365. The period-doubled solution is modulated but with fewer oscillations per complete cycle and smaller amplitude than in Figure 27.

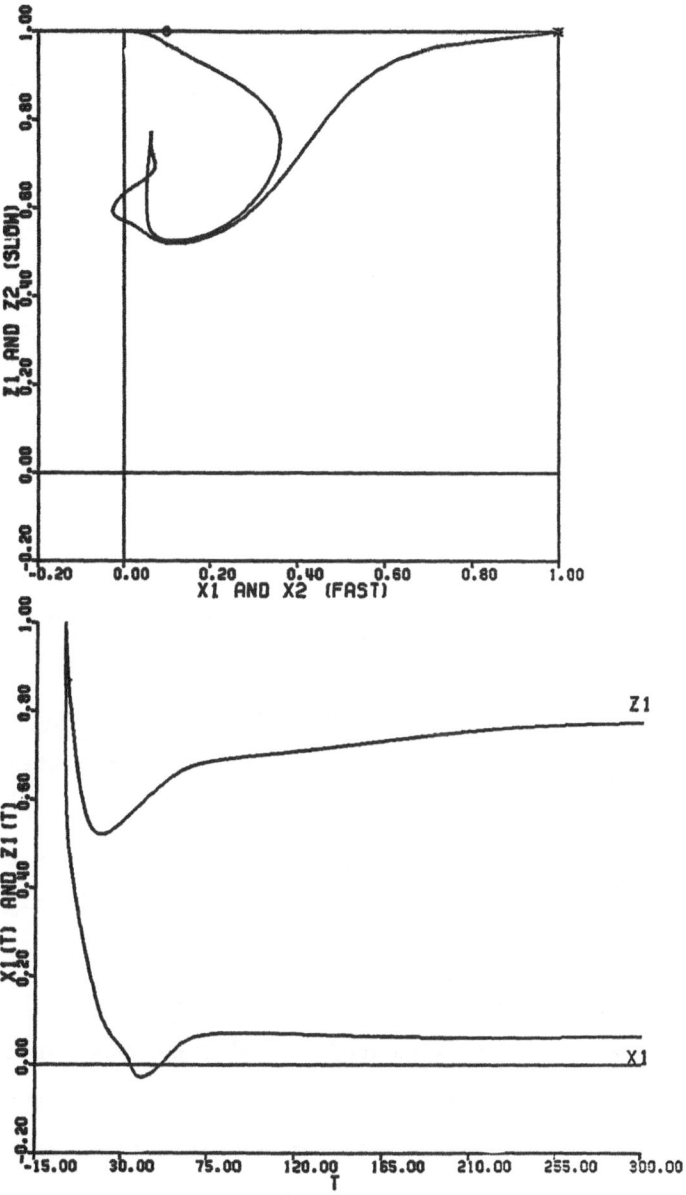

Figure 29: C_7 = .37. A small amplitude periodic solution near the
diagonal.

such a way that every other peak gradually becomes smaller than its neighboring peaks. Consequently a period-doubled solution is a periodic solution in which a pair of large-then-small peaks repeats itself periodically through time. Thus, although a period-doubled solution is periodic it is not a _regular_ periodic solution, due to the alternation in peak sizes. Such an irregular periodic solution is said to be period-doubled because at the transitional value of C_7 where the regular periodic solution becomes an irregular periodic solution, the period of the solution doubles.

A singular perturbation analysis has been developed to prove the existence of the regular periodic and period-doubled solutions as well as the transition between solutions (Carpenter and Grossberg, 1982b). This type of dynamic period doubling is different from the period doubling discovered by Feigenbaum (1978, 1979), which is currently a topic of great interest (Ruelle, 1981).

The existence of period-doubled solutions immediately suggests the possibility that chaotic solutions can also be generated by a gated pacemaker. Our numerical studies are insufficient to decide this issue, but we have observed a bifurcation of the period-doubled solutions to a new type of irregular periodic solution. In Figure 27, where $C_7 = .355$, the period-doubled amplitudes are slowly modulated on a time scale that is long compared to the interpeak duration. In terms of biological clocks, these slow modulations are suggestive of biorhythms and other slowly varying modulations of mood or activity level. In particular, if each peak represents a day, then the modulation in Figure 27 is on the order of magnitude of a month.

As C_7 increases to .365 (Figure 28) solution peaks are still slowly modulated, but all the peak amplitudes are smaller and there are fewer oscillations per cycle. When $C_7 = .37$ (Figure 29), the oscillations lie close to the diagonal. Finally, for sufficiently large choices of C_7, the oscillations are quenched and the variables

approach limits on the diagonal.

III.7. Metabolic Feedback and Slow Gain Control

The complete model is augmented with two types of slowly varying

feedback. The feedback that measures activity-generated metabolic

feedback is used to explain phenomena such as split rhythms, whereas

the feedback due to slow gain control is used to explain phenomena

such as long-term after-effects. Only the model of a diurnal animal

will be considered herein. As before, the nocturnal model is identical

except that $J(t)$ appears in the x_2 instead of the x_1 conductance.

To define metabolic feedback, or fatigue, we suppose that the on-

cell potential $x_1(t)$ generates a signal $G(x_1(t))$ that energizes obser-

vable activity, such as wheel-turning. We typically choose a sigmoid

output function

$$G(w) = \begin{cases} \dfrac{w^2}{c_8^2+w^2} & \text{if } w > 0 \\ \\ 0 & \text{if } w \leq 0 \end{cases} \qquad , \qquad (65)$$

and assume that the model animal becomes active if $G(x_1(t))$ exceeds a

threshold C_9, and that suprathreshold activity is proportional to

$G(x_1(t))-C_9$. We assume that a metabolic debt increases linearly with

activity, and dissipates at a constant rate C_{10}. The metabolic feed-

back function is then

$$F(t) = C_{11} \int_0^t e^{-C_{10}(t-s)} \max[0,G(x_1(s))-C_9]ds \ . \qquad (66)$$

We assume that $F(t)$ directly excites the off-cell(s) v_2 and thereby

indirectly inhibits the on-cell(s) v_1 via competitive feedback. Equa-

tion (59) is then augmented to read

$$\frac{dx_2}{dt} = -x_2 + (1-x_2) [C_1 + C_2 f(x_2) z_2 + F] - (x_2 + C_3) C_4 g(x_1) . \qquad (67)$$

The metabolic feedback function $F(t)$ generates a split rhythm as follows. The build-up of $F(t)$ starts to inhibit the on-cell activity before it would otherwise be inhibited by the autonomous action of the pacemaker. The inhibition of on-cell activity, in turn, allows $F(t)$ to dissipate, thereby disinhibiting the on-cell before the end of its usual on-phase. This explanation tacitly assumes that the gate $z_1(t)$ has not become too depleted during this time. A similar sequence of events helps to explain the bimodal frequency of activity that is often observed before rhythm splitting occurs (Pittendrigh, 1974, p.449).

The gain control process can be interpreted in several ways. It is formally the same as a model of how cues become conditioned reinforcers by being associated with the activity of drive representations (Grossberg, 1982a). In the present model, the on-cell v_1 plays the role of a drive representation. The conditioning process is formally equivalent to a slowly changing gain that is a long-term average of the on-cell activity level. The conditioning term hereby buffers the system against short-term fluctuations in light and acts as a stabilizing parametric change in response to pervasive alterations in light patterning, say due to seasonal changes.

The inputs that are gated by the gain control process have several alternative physical interpretations that have not yet been theoretically or experimentally tested. One interpretation is that the input is light itself. Another interpretation is that reinforcing cues, such as a wheel that has been turned, are the input sources. Finally, many circadian properties still obtain if the inputs are chosen to be constant. In all these cases, the slow gain changes can be used to formally explain all the types of long-term after-effects

on the frequency of the free-running circadian rhythms that were described by Pittendrigh (1974, p. 441).

For example, to represent an input that is turned on when the
animal is awake we define

$$S(t) = \begin{cases} 1 \text{ if } G(x_1(t)) \geq \theta C_9 \\ 0 \text{ if } G(x_1(t)) < \theta C_9 \end{cases} . \tag{68}$$

The gain control function $y(t)$ is defined as a time-average of the
product of input $S(t)$ times the on-cell activity $x_1(t)$. Thus

$$\frac{dy}{dt} = -C_{12}y + C_{13}Sx_1 . \tag{69}$$

The function $y(t)$ in turn gates the input $S(t)$ to create a net input
to the on-cell v_1 that is proportional to $S(t)y(t)$, as in the following augmented equation for the on-cell potential:

$$\frac{dx_1}{dt} = -x_1 + (1-x_1)[C_1+J(t)+C_2f(x_1)z_1+Sy] - (x_1+C_3)C_4g(x_2) . \tag{70}$$

Note that two types of gating action occur in equation (70). Equations
(60), (61), and (65)-(70) complete the definition of a gated pacemaker
whose rhythm is modulated by both metabolic feedback and gain control
feedback.

CONCLUSION

Three concepts persistently occur throughout this article. One is the classical concept that a membrane equation can model fast electrical responses in cells. The second is the concept that mass action processes can be coupled to the membrane equation as conductance terms. The third is the concept that gating processes can be used to model the mass action dynamics of chemical transmitters. Our use of these concepts illustrates how a small number of simple mechanisms can generate a wide diversity of complex biological phenomena, as well as parametric experimental tests of the models that simulate these phenomena.

REFERENCES

J. Aschoff, 1960, Exogenous and endogenous components in circadian rhythms, Cold Spring Harbor Symp. Quant. Biol. 25, 11-26.

J. Aschoff, 1979, Influences of internal and external factors on the period measured in constant conditions, Z. Tierpsychol. 49, 225-249.

J. Atkinson and A. Ward, 1964, Intracellular studies of cortical neurons in chronic epileptogenic foci in the monkey, Exp. Neurol. 10, 285-295.

J.L. Barker and T.G. Smith, Jr., 1978, Electrophysiological studies of molluscan neurons generating bursting pacemaker potential activity, in: Abnormal Neuronal Discharges (M. Boisson and N. Chalazonitis, Editors), Springer-Verlag, New York, 359-387.

D.A. Baylor and A.L. Hodgkin, 1973, Detection and resolution of visual stimuli by turtle photoreceptors, J. Physiol. 234, 163-198.

D.A. Baylor and A.L. Hodgkin, 1974, Changes in time scale and sensitivity in turtle photoreceptors, J. Physiol. 242, 729-758.

D.A. Baylor, A.L. Hodgkin, and T.D. Lamb, 1974a, The electrical response of turtle cones to flashes and steps of light, J. Physiol. 242, 685-727.

D.A. Baylor, A.L. Hodgkin, and T.D. Lamb, 1974b, Reconstruction of the electrical responses of turtle cones to flashes and steps of light, J. Physiol. 242, 759-791.

D.A. Baylor, T.D. Lamb, and K.-W. Yau, 1979, The membrane current of single rod outer segments, J. Physiol. 288, 589-611.

P.R. Benjamin, 1978, Endogenous and synaptic factors affecting the bursting of double spiking molluscan neurosecretory neurons (Yellow Cells of Lymnae Stagnalis), in: Abnormal Neuronal Discharges (M. Boisson and N. Chalazonitis, Editors), Springer-Verlag, New York, 205-216.

S.A. Binkley, 1979, A timekeeping enzyme in the pineal gland, Scientific American 240, 66-71.

W.H. Calvin, 1972, Synaptic potential summation and repetitive firing mechanisms: Input-output theory for the recruitment of neurons into epileptic bursting firing patterns, Brain Res. 39, 71-94.

J.P. Card, J.N. Riley, and R.Y. Moore, 1980, The suprachiasmatic hypothalamic nucleus: Ultrastructure of relations to optic chiasm, Neurosci. Abstr. 6, 758.

G.A. Carpenter, 1977a, A geometric approach to singular perturbation problems with applications to nerve impulse equations, J. Differential Equations 23, 335-367.

G.A. Carpenter, 1977b, Periodic solutions of nerve impulse equations, J. Math. Analysis and Appl. 58, 152-173.

G.A. Carpenter, 1979, Bursting phenomena in excitable membranes, SIAM J. Appl. Math. 36, 334-372.

G.A. Carpenter, 1981, Normal and abnormal signal patterns in nerve cells, in: Mathematical Psychology and Psychophysiology (S. Grossberg, Editor), SIAM-AMS Proceedings 13, 49-90.

G.A. Carpenter and S. Grossberg, 1981, Adaptation and transmitter gating in vertebrate photoreceptors, J. Theor. Neurobiol. 1, 1-42.

G.A. Carpenter and S. Grossberg, 1982a, Circadian rhythms, chemical transmitters, and motivated behavior, submitted for publication.

G.A. Carpenter and S. Grossberg, 1982b, A neural theory of circadian rhythms amd related clocklike phenomena, submitted for publication.

S. Daan and C. Berde, 1978, Two coupled oscillators: Simulations of the circadian pacemaker in mammalian activity rhythms, J. Theor. Biol. 70, 297-313.

T. Deguchi, 1979, Circadian rhythm of serotonin N-acetyltransferase activity in organ culture of chicken pineal gland, Science 203, 1245-1247.

P.B. Detwiler, A.L. Hodgkin, and P.A. McNaughton, 1980, Temporal and spatial characteristics of the voltage response of rods in the retina of the snapping turtle, J. Physiol. 300, 213-250.

J.T. Enright, 1980, The Timing of Sleep and Wakefulness, Springer-Verlag, New York.

J.W. Evans, N. Fenichel, and J.A. Feroe, 1982, Double impulse solutions in nerve axon equations, SIAM J. Appl. Math. 42, 219-234.

D. Faber and M. Klee, 1972, Membrane characteristics of bursting pacemaker neurons in Aplysia, Nature 240, 29-31.

M.J. Feigenbaum, 1978, Quantitative universality for a class of nonlinear transformations, J. Statist. Phys. 19, 25-52.

M.J. Feigenbaum, 1979, The universal metric properties of nonlinear transformations, J. Statist. Phys. 21, 669-706.

J.A. Feroe, 1982, Existence and stability of multiple impulse solutions of a nerve equation, SIAM J. Appl. Math. 42, 235-246.

R. FitzHugh, 1961, Impulses and physiological states in theoretical models of nerve membrane, Biophysical J. 1, 445-466.

S. Grossberg, 1968, Some physiological and biochemical consequences of psychological postulates, Proc. Natl. Acad. Sci. 60, 758-765.

S. Grossberg, 1969, On the production and release of chemical transmitters and related topics in cellular control, J. Theor. Biol. 22, 325-364.

S. Grossberg, 1972a, A neural theory of punishment and avoidance, I. Qualitative theory, Math. Biosci. 15, 39-67.

S. Grossberg, 1972b, A neural theory of punishment and avoidance, II. Quantitative theory, Math. Biosci. 15, 253-285.

S. Grossberg, 1975, A neural model of attention, reinforcement, and discrimination learning, Inter. Rev. Neurobiol. 18, 263-327.

S. Grossberg, 1980, How does a brain build a cognitive code?, Psychol. Rev. 87, 1-51.

S. Grossberg, 1982a, The processing of expected and unexpected events during conditioning and attention: A psychophysiological theory, Psychol. Rev. 89, 529-572.

S. Grossberg, 1982b, Some psychophysiological and pharmacological correlates of a developmental, cognitive, and motivational theory, in: Cognition and Brain Activity (J. Cohen, R. Karrer, and P. Tueting, Editors), New York Academy of Sciences, New York.

S. Grossberg, 1982c, A psychophysiological theory of reinforcement, drive, motivation, and habit, J. Theor. Neurobiol., in press.

G.A. Groos and J. Hendriks, 1979, Regularly firing neurons in the rat suprachiasmatic nucleus, Experientia 35, 1597-1598.

G.A. Groos and R. Mason, 1978, Maintained discharge of rat suprachiasmatic neurons at different adaptation levels, Neurosci. Lett. 8, 59-64.

M.R. Guevara, L. Glass, and A. Shrier, 1981, Phase locking, period-doubling bifurcations, and irregular dynamics in periodically stimulated cardiac cells, Science 214, 1350-1353.

E. Gwinner, 1974, Testosterone induces "splitting" of circadian locomotor activity rhythms in birds, Science 185, 72-74.

S.P. Hastings, 1982, Single and multiple pulse waves for the FitzHugh-Nagumo equations, SIAM J. Appl. Math. 42, 247-260.

A.L. Hodgkin and A.F. Huxley, 1952, A quantitative description of membrane current and its application to conduction and excitation in nerve, J. Physiol. 117, 500-544.

K. Hoffmann, 1971, Splitting of the circadian rhythm as a function of light intensity, in: Biochronometry (M. Menaker, Editor), National Academy of Sciences, Washington, D.C., 134-150.

M. Jouvet, 1974, Monoaminergic regulation of the sleep-waking cycle in the cat, in: Circadian Oscillations and Organization in Nervous Systems (C.S. Pittendrigh, Editor), M.I.T. Press, Cambridge, Mass., 499-508.

M. Jouvet, J. Mouret, G. Chouvet, and M. Siffre, 1974, Toward a 48-hour day: Experimental bicircadian rhythm in man, in: Circadian Oscillations and Organization in Nervous Systems (C.S. Pittendrigh, Editor), M.I.T. Press, Cambridge, Mass., 491-497.

E. Kandel and W. Spencer, 1961, Electrophysiology of hippocampal neurons, II. After-potentials and repetitive firing, J. Neurophysiol. 24, 243-259.

M. Kawato and R. Suzuki, 1980, Two coupled neural oscillators as a model of the circadian pacemaker, J. Theor. Biol. 86, 547-575.

K.R. Kramm, 1971, Circadian activity in the antelope ground squirrel, Ammospermophilus leucurus, Ph.D. Thesis, Univ. of California, Irvine.

R.E. Kronauer, C.A. Czeisler, S.F. Pilato, M.C. Moore-Ede, and E.D. Weitzman, 1982, Mathematical model of the human circadian system with two interacting oscillators, Amer. J. Physiol. 242, R3-R17.

E. Lábos and E. Láng, 1978, On the behavior of snail (Helix pomatia) neurons in the presence of cocaine, in: Abnormal Neuronal Discharges (M. Boisson and N. Chalazonitis, Editors), Springer-Verlag, New York, 177-188.

A.J. Lewy, T.A. Wehr, F.K. Goodwin, D.A. Newsome, and S.P. Markey, 1980, Light suppresses melatonin secretion in humans, Science 210, 1267-1269.

T.-Y. Li and J.A. Yorke, 1975, Period three implies chaos, Amer. Math. Mon. 82, 985-992.

D.W. Lincoln, J. Church, and C.A. Mason, 1975, Electrophysiological activation of suprachiasmatic neurones by changes in retinal illumination, Acta Endocrinol. (suppl., Kbh.) 199, 184.

M. Markowitz, L. Rotkin, and J.F. Rosen, 1981, Circadian rhythms of blood minerals in humans, Science 213, 672-674.

R.M. May, 1976, Simple mathematical models with very complicated dynamics, Nature 261, 459-467.

H.P. McKean, Jr., 1970, Nagumo's equation, Adv. Math. 4, 209-223.

P.J. Mill, 1977, Ventilation motor mechanisms in the dragonfly and other insects, in: Identified Neurons and Behavior of Arthropods (G. Hoyle, Editor), Plenum Press, New York, 187-208.

R.Y. Moore, 1973, Retinohypothalamic projection in mammals: A comparative study, Brain Res. 49, 403-409.

R.Y. Moore, 1974, Visual pathways and the central neural control of diurnal rhythms, in: Circadian Oscillations and Organization in Nervous Systems (C.S. Pittendrigh, Editor), M.I.T. Press, Cambridge, Mass., 537-542.

R.Y. Moore, J.P. Card, and J.N. Riley, 1980, The suprachiasmatic hypothalamic nucleus: Neuronal ultrastructure, Neurosci. Abstr. 6, 758.

M.C. Moore-Ede, F.M. Sulzman, and C.A. Fuller, 1982, The Clocks That Time Us, Harvard University Press, Cambridge, Mass.

J. Nagumo, S. Arimoto, and S. Yoshizawa, 1962, An active pulse transmission line simulating nerve axon, Proceedings IEEE 50, 2061-2070.

E.J. Nestler, M. Zatz, and P. Greengard, 1982, A diurnal rhythm in pineal protein 1 content mediated by β-adrenergic neurotransmission, Science 217, 357-359.

H. Nishino, K. Koizumi, and C.M. Brooks, 1976, The role of suprachiasmatic nuclei of the hypothalamus in the production of circadian rhythm, Brain Res. 112, 45-59.

B.J. Nunn, G.G. Matthews, and D.A. Baylor, 1980, Comparison of voltage and current responses of retinal rod photoreceptors, Fed. Proc. 39, 2066.

P. Passouant and I. Oswald (Editors), 1979, Pharmacology of the States of Alertness, Pergamon Press, New York.

G.E. Pickard and F.W. Turek, 1982, Splitting of the circadian rhythm of activity is abolished by unilateral lesions of the suprachiasmatic nuclei, Science 215, 1119-1121.

C.S. Pittendrigh, 1960, Circadian rhythms and the circadian organization of living systems, Cold Spring Harbor Symp. Quant. Biol. 25, 159-182.

C.S. Pittendrigh, 1974, Circadian oscillations in cells and the circadian organization of multicellular systems, in: Circadian Oscillations and Organization in Nervous Systems (C.S. Pittendrigh, Editor), M.I.T. Press, Cambridge, Mass., 437-458.

C.S. Pittendrigh and S. Daan, 1976, A functional analysis of circadian pacemakers in nocturnal rodents, I. The stability and lability of spontaneous frequency, J. Comp. Physiol. 106, 223-252.

R.E. Plant and M. Kim, 1975, On the mechanism underlying bursting in the Aplysia abdominal ganglion R15 cell, Math Biosci. 26, 357-375.

J. Rinzel and J.B. Keller, 1973, Traveling wave solutions of a nerve conduction equation, Biophysical J. 13, 1313-1337.

F.A. Roberge, M. Gulrajani, H.H. Jasper, and P.A. Mathieu, 1978, Ionic mechanisms for rhythmic activity and bursting in nerve cells, in: Abnormal Neuronal Discharges (M. Boisson and N. Chalazonitis, Editors), Springer-Verlag, New York, 389-405.

D. Ruelle, 1981, Differentiable dynamical systems and the problem of turbulence, Bull. Amer. Math. Soc. 5, 29-42.

D.F. Russell and D.K. Hartline, 1978, Bursting neural networks: A re-examination, Science 200, 453-456.

J.S. Takahashi, H. Hamm, and M. Menaker, 1980, Circadian rhythms of melatonin release from individual superfused chicken pineal glands in vitro, Proc. Natl. Acad. Sci. USA 77, 2319-2322.

W.N. Tapp and F.A. Holloway, 1981, Phase shifting circadian rhythms produces retrograde amnesia, Science 211, 1056-1058.

R.F. Thompson, 1967, Foundations of Physiological Psychology, Harper and Row, New York.

A. Ward, Jr., 1969, The epileptic neuron: Chronic foci in animals and man, in: Basic Mechanisms of the Eplepsis, Little, Brown, and Company, Boston, 263-288.

T.A. Wehr and A. Wirz-Justice, 1982, Circadian rhythm mechanisms in affective illness and in antidepressant drug action, Pharmacopsychiat. 15, 30-38.

R. Wever, 1962, Zum Mechanismus der biologischen 24-Stunden-Periodik, Kybernetik 1, 139-154.

R. Wever, 1975, The circadian multi-oscillator system of man, <u>Inter. J. Chronobiol.</u> 3, 19-55.

R.A. Wever, 1979, <u>The Circadian System of Man: Results of Experiments Under Temporal Isolation</u>, Springer-Verlag, New York.

A.T. Winfree, 1980, <u>The Geometry of Biological Time</u>, Springer-Verlag, New York.

K.-W. Yau, P.A. McNaughton, and A.L. Hodgkin, 1981, Effect of ions on the light-sensitive current in retinal rods, <u>Nature</u> 292, 502-505.

M. Zatz and M.J. Brownstein, 1979, Intraventricular carbachol mimics the effect of light on the circadian rhythm in the rat pineal gland, <u>Science</u> 203, 358-361.

Bio-mathematics

Managing Editor: S. A. Levin

Springer-Verlag
Berlin
Heidelberg
New York

Volume 8
A. T. Winfree

The Geometry of Biological Time

1979. 290 figures. XIV, 530 pages
ISBN 3-540-09373-7

The widespread appearance of periodic patterns
in nature reveals that many living organisms are
communities of biological clocks. This land-
mark text investigates, and explains in mathematical terms, periodic processes in living
systems and in their non-living analogues. Its
lively presentation (including many drawings),
timely perspective and unique bibliography will
make it rewarding reading for students and re-
searchers in many disciplines.

Volume 9
W. J. Ewens

Mathematical Population Genetics

1979. 4 figures, 17 tables. XII, 325 pages
ISBN 3-540-09577-2

This graduate level monograph considers the
mathematical theory of population genetics,
emphasizing aspects relevant to evolutionary
studies. It contains a definitive and comprehensive discussion of relevant areas with references
to the essential literature. The sound presentation and excellent exposition make this book a
standard for population geneticists interested in
the mathematical foundations of their subject
as well as for mathematicians involved with
genetic evolutionary processes.

Volume 10
A. Okubo

Diffusion and Ecological Problems:
Mathematical Models

1980. 114 figures, 6 tables. XIII, 254 pages
ISBN 3-540-09620-5

This is the first comprehensive book on mathematical models of diffusion in an ecological
context. Directed towards applied mathematicians, physicists and biologists, it gives a
sound, biologically oriented treatment of the
mathematics and physics of diffusion.

Journal of Mathematical Biology

ISSN 0303-6812   Title No. 285

Editorial Board:
H.T.Banks, Providence, RI; **H.J.Bremermann,** Berkeley,
CA; **J.D.Cowan,** Chicago, IL; **J.Gani,** Lexington, KY;
K.P.Hadeler (Managing Editor), Tübingen;
F.C.Hoppensteadt, Salt Lake City, UT; **S.A.Levin**
(Managing Editor), Ithaca, NY; **D.Ludwig,** Vancouver; .
L.A.Segel, Rehovot; **D.Varjú,** Tübingen in cooperation
with a distinguished advisory board.

The **Journal of Mathematical Biology** publishes papers in
which mathematics leads to a better understanding of biological phenomena, mathematical papers inspired by biological research and papers which yield new experimental data
bearing on mathematical models. The scope is broad, both
mathematically and biologically and extends to relevant
interfaces with medicine, chemistry, physics, and sociology.
The editors aim to reach an audience of both mathematicians
and biologists.

Contents:

Subscription information and sample copy upon request

Springer-Verlag
Berlin
Heidelberg
New York

Lecture Notes in Biomathematics